STAT101

Statistics Software for Today's Students

Christine Siegel, Ph. D.

Limited Warranty

Addison-Wesley warrants that STAT101 ("the program") will substantially conform to the published specifications and to the documentation during the period of 90 days from the date of original purchase, provided that it is used on the computer hardware and with the operating system for which it was designed. Addison-Wesley also warrants that the magnetic media on which the program is distributed and the documentation are free from defects in materials and workmanship during the period of 90 days from the date of original purchase. Addison-Wesley will replace defective media or documentation or correct substantial program errors at no charge, provided you return the item with dated proof of purchase to Addison-Wesley within 90 days of the date of original purchase. If Addison-Wesley is unable to replace defective media or documentation or correct substantial program errors, your license fee will be refunded. These are your sole remedies for any breach of warranty.

Except as specifically provided above, Addison-Wesley makes no warranty or representation, either express or implied, with respect to this program, documentation, or media, including their quality, performance, merchantability, or fitness for a particular purpose.

Because programs are inherently complex and may not be completely free of errors, you are advised to verify your work. **In no event will Addison-Wesley or Minitab be liable for direct, indirect, special, incidental, or consequential damages arising out of the use of or inability to use the program, documentation, or media,** even if advised of the possibility of such damages. Specifically, neither Addison-Wesley nor Minitab is responsible for any costs including, but not limited to, those incurred as a result of lost profits or revenue (in any case, the program must be used only for educational purposes, as required by your license), loss of use of the computer program, loss of data, the costs of recovering such programs or data, the cost of any substitute program, claims by third parties, or for other similar costs. In no case shall the liability of Addison-Wesley or Minitab exceed the amount of the license fee.

The warranty and remedies set forth above are exclusive and in lieu of all others, oral or written, express or implied. No Addison-Wesley dealer, distributor, agent, or employee is authorized to make any modification or addition to this warranty.

Some statutes do not allow the exclusion of implied warranties; if any implied warranties are found to exist, they are hereby limited in duration to the 90-day life of the express warranties given above. Some states do not allow the exclusion or limitation of incidental or consequential damages, nor any limitation on how long implied warranties last, so these limitations may not apply to you. This warranty gives you specific legal rights, and you may also have other rights which vary from state to state.

To obtain performance of this warranty, return the item with dated proof of purchase within 90 days of the purchase date to: Addison-Wesley Publishing Company, Inc., STAT101—Marketing Department, Jacob Way, Reading, MA 01867.

STAT101

Statistics Software for Today's Students

Addison-Wesley • Minitab Inc.

Addison-Wesley Publishing Company

Reading, Massachusetts • Menlo Park, California • New York
Don Mills, Ontario • Wokingham, England • Amsterdam • Bonn
Sydney • Singapore • Tokyo • Madrid • San Juan • Milan • Paris

STAT101 is published by Addison-Wesley Publishing Company, Inc.
Contributors include:

Executive Editor: Michael Payne
Managing Editor: Kazia Navas
Development Editor: Joan Carey
Software/Special Projects Manager: Mary Coffey
Production Coordinator: Barbara Ames
Editorial Assistant: Maureen Lawson
Prepress Consultant: John Webber
Manufacturing Manager: Trish Gordon
Cover Design: Peter Blaiwas
Compositor: Jamie James

The installation program used to install *STAT101*, INSTALL, is based on licensed software provided by Knowledge Dynamics Corp., P.O. Box 1558, Canyon Lake, Texas 78130-1558 (USA). INSTALL is Copyright © 1987-1991 by Knowledge Dynamics Corp., which reserves all copyright protection worldwide. INSTALL is provided to you for the exclusive purpose of installing *STAT101*. Minitab Inc. has made modifications to the software as provided by Knowledge Dynamics Corp., and thus the performance and behavior of the INSTALL program shipped with *STAT101* may not represent the performance and behavior of INSTALL as shipped by Knowledge Dynamics Corp. Addison-Wesley is exclusively responsible for the support of *STAT101*, including support during the installation phase. In no event will Knowledge Dynamics Corp. be able to provide any technical support for *STAT101*.

MINITAB is a registered trademark of Minitab Inc.

STAT101 is a registered trademark of Minitab Inc.

IBM is a registered trademark, and PC-DOS is a trademark of International Business Machines Corporation.

80286 is a registered trademark of Intel Corporation.

Mac and Macintosh are registered trademarks of Apple Computer, Inc.

Lotus, 1-2-3, and Symphony are registered trademarks of Lotus Development Corporation.

General Notice: Some of the product names used herein have been used for identification purposes only and may be trademarks of their respective companies.

Chapters 1, 2, 4, 5, and Appendixes A and C are copyrighted © 1994 by Minitab Inc. All rights reserved.

Chapters 3, 6, and Appendix B are copyrighted © 1994 by Addison-Wesley Publishing Company, Inc.

It is a violation of copyright law to make a copy of the accompanying software except for backup purposes to guard against accidental loss or damage. Addison-Wesley assumes no responsibility for errors arising from duplication of the original programs.

All rights reserved. No part of this publication may be reproduced, stored in a retrieval system, or transmitted, in any form or by any means, electronic, mechanical, photocopying, recording, or otherwise, without the prior written permission of the publisher. Printed in the United States of America.

ISBN 0-201-59087-5 (5 1/4" package)
      0-201-59088-3 (3 1/2" package)
      0-201-93792-1 (manual)

1 2 3 4 5 6 7 8 9 10-AL-96959493

Preface

Welcome

Welcome to STAT101.

STAT101 is an inexpensive statistics software package that provides a wide range of data analysis capabilities. You will find it a useful tool that can handle all introductory-level analysis and data description tasks.

STAT101 is produced by Minitab Inc., long recognized as a leading developer of reliable and easy-to-use statistical software. If you intend to continue your study of statistics, you will make a smooth transition to the commercial package; the structure, style, commands, and file formats are fully compatible. More college students are trained on MINITAB than any other statistical software package, and STAT101 is the perfect first step for beginners.

Objectives

STAT101 is a data analysis tool for students in introductory statistics courses (or comparable mathematics courses). The primary objectives of this package are to

- provide a complete but inexpensive data analysis package that you can use for entry-level courses
- give you experience using statistical software that you can apply to using MINITAB if you need a more powerful statistical package in your continuing studies
- give you software that lets you use statistical techniques quickly and efficiently that would normally involve too many calculations by hand
- provide real-life data sets as samples for learning statistical techniques

Manual

This manual has six chapters and four appendixes that provide the information you need to install and use STAT101. The first five chapters are task-oriented, and you can follow along with the examples on your own computer because they are drawn from data sets that come with the STAT101 package. Chapter 6, the Command Reference, is organized by command. Most commands include examples that you can duplicate from start to finish.

Chapter 1. Getting Started

This chapter provides a summary of the STAT101 package, installation instructions, and an explanation of manual conventions and basic procedures, including how to stop and start STAT101.

Chapter 2. Sample Sessions

The two sample sessions (Basic and Advanced) give a hands-on introduction to STAT101 Statistics Software for Today's Students. You assume the role of data analyst as you examine data on the exercise patterns and physical condition of college students.

Chapter 3. Using the Data Editor

This chapter introduces you to STAT101's Data Editor, including entering and editing data, using the Data Editor menu, and getting Data Editor online help. This chapter can be used as a tutorial, where you proceed through the information step-by-step on your computer. Reference information on stored constants and data formats (numeric and alpha) is also included, as well as a summary of all Data Editor keyboard functions.

Chapter 4. Issuing Commands

This chapter introduces STAT101's command structure, including how to issue STAT101 commands and subcommands, getting online help for STAT101 commands, and dealing with special STAT101 symbols and prompts. This chapter can be used as a tutorial, where you proceed through the information step-by-step on your computer.

Chapter 5. Handling Files and Printing

This chapter shows you how to save and retrieve STAT101 worksheets (including those in the sample data sets that accompany the software), get data from other applications in and out of STAT101, create an outfile to save output and commands, delete files, and finally, how to print hard copies of your work.

Chapter 6. Command Reference

This chapter begins with a table of command groups, organized logically, so you can locate a task and then find the commands associated with that task. The bulk of this chapter is a complete listing, organized alphabetically, of each STAT101 command, including command and subcommand syntax, information on the format, options, and operation of each command, and an example of how to use it with data sets included with the STAT101 software.

Appendix A: Troubleshooting

Troubleshooting alerts you to common error messages that STAT101 displays when it detects error conditions, explaining the message, likely causes of the error, and possible corrective actions.

Appendix B: Sample Data Sets

Sample Data Sets describes each data set that comes with the STAT101 software, including basic information (column, name, count, description) on each variable in the file.

Appendix C: Quick Reference

Quick Reference lists all STAT101 commands and their syntax and use.

Appendix D: MINITAB Software Products

MINITAB Software Products lists other statistical software products and additional documentation developed by Minitab Inc., all of which are fully compatible with STAT101.

Features

STAT101 features a storage area of 100 columns and 100 constants, allowing 2000 data points. Using STAT101, you can

- enter, save, and retrieve numeric and alphabetic data
- edit, manipulate, and transform data, using commands or the spreadsheet-like Data Editor that makes handling data easy
- describe data in tabular and graphical format (including scatter plots, histograms, and tables)
- generate random data from any of the built-in statistical distributions
- calculate p-values from continuous and discrete distribution functions
- analyze data, using a variety of statistical procedures
- summarize data with descriptive statistics
- perform simple and multiple regression

- perform one- and two-way analysis of variance
- perform nonparametric analysis
- perform time-series analysis
- create and execute macros—STAT101 programs you can use to automate repetitive tasks or to create your own customized features

Technical Support

Neither Addison-Wesley nor Minitab Inc. provides phone assistance to students using *STAT101- Statistics Software for Today's Students*. Phone assistance is, however, provided to **registered** instructors who adopt *STAT101 - Statistics Software for Today's Students*.

If you encounter difficulty using the *STAT101* software,

- Consult the Reference section of the User's Manual for information on the commands or procedures you are trying to perform.
- Use the Help screens to locate specific program or error message information.

If you need to ask your instructor for assistance, describe your question or problem in detail. Be sure to note what you were doing (the steps or procedures you followed) when the problem occurred. Also write down the exact error message (if any).

Contents

Chapter 1: Getting Started 1
   Before You Begin 1
   Installation 4
      Making Backup Copies of STAT101 9
      Starting and Stopping STAT101 10

Chapter 2: Sample Sessions 13
   How to Use the Sample Sessions 13
   Basic Sample Session 14
   Advanced Sample Session 28

Chapter 3: Using the Data Editor 39
   Objectives 39
   Opening the Data Editor 39
   Moving Around 41
   Getting Help 43
   Entering Data 44
   Editing the Worksheet 46
   Naming Columns 47
   Stored Constants 49
   Data Formats 49
   Data Editor Keyboard Summary 51

Chapter 4: Issuing Commands 53
   Objectives 53
   Commands and Arguments 53
   Learning STAT101's Command Syntax 54
   Using Subcommands 55

    Typographical Conventions for Commands   56

    Syntax Conventions   57

    Using Help   58

    Symbols   58

    STAT101 Prompts   59

Chapter 5: Handling Files and Printing   61

    Objectives   61

    File Types   61

    Saving and Retrieving STAT101 Worksheets   62

    Saving STAT101 Worksheet Data in a Text (ASCII) File   64

    Saving Your STAT101 Session   67

    Deleting Files   67

    Printing   68

    STAT101 File Summary   70

Chapter 6: Command Reference   71

    Command Groups   71

    Command Syntax   82

Appendix A: Troubleshooting   221

    Not Reading Drive   221

    Can't Find File   222

    File Name Problem   222

    Can't Start STAT101   223

Appendix B: Sample Data Sets   225

Appendix C: Quick Reference   273

Appendix D: MINITAB Software Products   283

    Documentation   283

    Software   284

Index   287

1

Getting Started

Before You Begin

Checking Your Package

The STAT101 package should contain the following items:

- One STAT101 Program disk (high-density), either 5 1/4" or 3 1/2".
- This manual.
- Warranty Registration Card. Fill out this card and mail it back to Addison-Wesley Publishing Company. This registers you as a licensee of the package, and makes you eligible to receive further product information.
- License Agreement that spells out the terms of the agreement you make as a user with Addison-Wesley when you purchase STAT101.

System Requirements

You will need the following computer equipment to run STAT101:

- An IBM PC-AT or compatible computer with the INTEL 80286 processor or higher
- A hard disk with at least 2M of available storage space
- A floppy disk drive that accepts high-density diskettes
- At least 460K of available conventional memory (RAM)
- DOS release 3.3 or higher

A math coprocessor will make STAT101 run faster, but it is not required. If you have a color monitor you can configure your own colors, using MSETUP, but a monochrome monitor is sufficient (80-character width).

STAT101 does not require a graphics card or a special printer device driver. If you are running Microsoft Windows, STAT101 runs as a non-Windows application under Version 3.1. It does not support a mouse.

Typographical Conventions

You control STAT101 by keystrokes you enter from the keyboard of your computer. Throughout this manual you will find step-by-step instructions for installing and running STAT101. All instructions are preceded by a bullet (•), including things you type, which are in **boldface**. For example:

- Type **HISTOGRAM C1** after the STAT> prompt.

Although this manual shows commands and file names in uppercase, you can use any combination of upper- and lowercase; STAT101 doesn't distinguish between them.

Key combinations are indicated by a + between the two keys, like [Ctrl] + [Enter]. When you see a key combination, press and hold down the first key and, while it is still depressed, press the second key.

The symbol [Enter] refers to the Enter key. Some keyboards show this key as [Return] or [←Return].

Technical Support

Addison-Wesley Publishing Co., Inc. provides technical support for registered instructors who have adopted STAT101 for their classes.

If you have trouble using STAT101, look for help in the following places:

- This user's manual.
- Online help, either from the Session screen or Data Editor (the Data Editor has a special Troubleshooting topic).
- Appendix A, Troubleshooting.
- The file READ.ME that was installed with STAT101. The READ.ME file contains a listing of known STAT101 problems and includes strategies for working around them. See Appendix A for information on accessing the READ.ME file.

If you cannot resolve the problem using these sources, write down the following information about the computer on which you are working: brand, model, operating system version, number and type of disk drives, available memory, and whether or not a math coprocessor is installed. Write down the exact steps you followed and any error messages you received, and then go to your instructor, who has access to STAT101 technical support.

Using DOS

DOS is an acronym for Disk Operating System. A computer's operating system is a program that controls the operation of computer programs and data. If your computer didn't have an operating system on its hard disk, nothing would appear on the screen after you turn on your computer.

Your computer has one or two disk drives that read floppy disks. The first floppy disk drive that lights up after you turn on the computer is the A: drive. If you have a second floppy drive, it is called the B: drive; the hard drive is usually called the C: drive.

If you have more than one hard drive, the next one is called D:, and then E:, and so on.

When you turn on your computer, the first screen you see depends on what version of DOS you are running and whether you are running Windows or a menu application. This manual assumes you are operating from the DOS *command prompt,* usually seen as C:\> or C>, shown in Fig. 1–1.

(This manual uses an arrow (⇨) or figure number to mark a section that shows what your computer screen should look like.)

FIGURE 1–1

```
Microsoft (R) MS-DOS (R) Version 5.00
                (C)Copyright Microsoft Corp. 1981-1991

C:\>
```

Your version of DOS may be different from that shown in Fig. 1–1.

This prompt has two parts: the first, C:, indicates the active drive. The second identifies the active *directory*. A directory is like a file folder in which you store information, and is always preceded by a backslash (\). For example, the DOS prompt you will see later, C:\STAT101>, indicates that drive C: is active and that you are in the STAT101 directory of the disk in that drive. The command prompt C:\> indicates that you are in the home or *root directory* on the hard drive C, because there is no directory indicated after the \. The arrow > marks the end of the prompt. DOS uses this prompt to ask you for input; when the prompt appears, you type commands after it.

Installation

Because the original STAT101 diskette contains compressed files, you must use STAT101's Install utility to install the files on your hard disk. If you merely copy the files on the program disk directly to your hard disk, you will not be able to use STAT101.

Installing STAT101

The STAT101 Install utility automatically installs the STAT101 program into a directory on your hard disk that it names STAT101.

To install STAT101:

- Make sure your STAT101 diskette is not write-protected. For 3 1/2" diskettes, the small window in the upper-right corner of the diskette should be closed; for 5 1/4" diskettes, make sure the notch on the right edge of the diskette is visible and not covered by a sticker
- If necessary, start your computer and turn on the monitor.

The DOS prompt will appear, unless your computer has a shell menu system over DOS, in which case you should

- Go to the DOS prompt (usually C:\> or C>).

If you are running Microsoft Windows, double-click the MS-DOS program icon in the Main window of the Program Manager.

- Insert the STAT101 diskette into drive A or B, depending on your diskette and drive size.
- Type **A: INSTALL** or **B: INSTALL,** whichever is appropriate for your drive.

⇨ C:\> A:INSTALL

- Press [Enter].

Personalization

If this is the first time you are installing STAT101, after a few seconds you will see *personalization* instructions on your screen, asking you to enter your name. Personalizing a disk identifies it as yours. You do this step only one time, but you must complete it in order to install STAT101 properly.

- Type your name after the arrow.

If you make a mistake, press the backspace ←Backspace key (← on some keyboards) to delete what you've typed, and then start over.

- Press Enter when you are done typing your name.

STAT101 displays what you have typed, and asks if your name is correct.

- Press Y for Yes if it is correct, N for No if it is not, or X to exit the installation procedure.

If you say no, STAT101 gives you the opportunity to correct your name. When your name is correct and you have verified it by pressing Y, STAT101 personalizes your disk and then proceeds to the Installation screen.

- Press Enter to continue.

INSTALL Main Menu

The INSTALL Main Menu appears on the screen shown in Fig. 1–2.

FIGURE 1–2

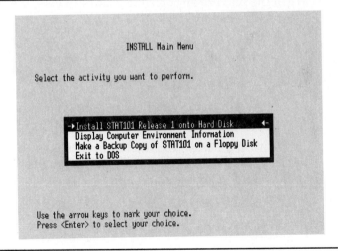

The Main Menu lists four options, highlighting the first.

The highlighted option, the one preselected by the program, is called the *default*. Most users can accept the default menu selections throughout the installation process. If you don't want to accept the default option, press ↑ or ↓ to highlight the option you want to accept, and then press Enter.

- Press [Enter] to choose Install STAT101 Release 1 onto Hard Disk (the default menu option).

The next screen provides a menu choice of the available disk drives on your computer where you can install STAT101; Drive C: will be highlighted as the default.

- Press [Enter] to install STAT101 on the C drive, or select a different disk drive.

The next screen recommends installing the software in the STAT101 directory.

- Press [Enter] to install STAT101 in the STAT101 directory, or specify a different directory by pressing [Bksp] to delete the characters, typing in a new name, and finally pressing [Enter].

The Install utility then checks your hard disk to make sure it has enough space. If your disk is full, Install alerts you and recommends that you make room by deleting files.

- Follow the directions on the screen until STAT101 is installed.

When the files are all copied to your hard disk, the Install utility asks if it can modify your AUTOEXEC.BAT file.

- Press [Y] if you want to be able to start STAT101 from any directory on your hard disk, and follow the directions on the screen.

Install then asks if it can modify your CONFIG.SYS file.

- Press [Y] so that STAT101 can run at optimal performance, and then follow the directions on the screen.

Install then informs you that the next step involves changing screen settings, using the MSETUP program (required for a monochrome monitor, optional for a color monitor).

To start MSETUP from the end of the Install program:

- Press [Enter].

The MSETUP Main Menu appears.

MSETUP

The MSETUP program lets you change screen colors and Data Editor settings from their default settings. You can change the selections you make in MSETUP at any time.

If you have a monochrome monitor you must use MSETUP to select the appropriate screen settings. If you have a color monitor, you don't have to do anything with MSETUP, although you can change colors if you want to.

If you just installed STAT101, the MSETUP menu is already on your screen, and you can skip the next section and proceed to Using MSETUP.

Opening the MSETUP Menu

If the MSETUP menu is not on your screen and you want to modify screen colors or Data Editor settings, you need to make STAT101 the current directory.

If the prompt is not C:\STAT101, perform the following steps:

- Type **CD \STAT101** and press Enter.

(CD stands for Change Directory.) The prompt appears now as C:\STAT101>.

- Type **MSETUP** and press Enter.

The MSETUP menu appears.

Using MSETUP

When you start MSETUP, the MSETUP Main Menu appears, as shown in Fig. 1–3.

FIGURE 1–3

```
┌──────────────── MSETUP Main Menu ────────────────┐
│ MSETUP Help                                       │
│ General Screen Colors                             │
│ Data Editor Parameters                            │
│ Set Default Colors for Color Monitor              │
│ Set Default Colors for Monochrome Monitor         │
└──────────────────<ESC> to exit───────────────────┘
```

The first option, MSETUP Help, is highlighted. It gives instructions on navigating around MSETUP.

STAT101 is installed with default color settings for color monitors. If you have a monochrome monitor, you should select the monochrome default if you want the screen elements to be most visible. To select the monochrome default:

Installation 7

- If you have a monochrome monitor, press ↓ four times to highlight the last option, Set Default Colors for Monochrome Monitor.
- Press Enter.

STAT101 automatically adjusts its color settings to display best on a monochrome monitor. You can, however, select different settings for both color and monochrome monitors. Obviously, your choices will be more limited for monochrome monitors.

To select different display colors:

- Highlight General Screen Colors on the MSETUP Main Menu and press Enter. (If you are using the arrows on your key pad be sure the Num Lock key is turned off.)
- Use the arrow keys to highlight the cell that shows the color combination you want for background and character colors.
- Press Enter to save the color combination you chose and return to the MSETUP Main Menu.

You can also change the settings of the Data Editor screen (the place that displays the columns and rows of your data) for both color and column width:

- Highlight Data Editor Parameters on the MSETUP Main Menu and press Enter.
- Use the arrow keys to highlight the settings you want to change, pressing Enter once you've highlighted the setting and Esc to return to the MSETUP Main Menu.

If you decide against any of the color choices you've made, you can always reset the screen to the defaults by selecting the appropriate default command from the MSETUP menu, depending on whether your monitor is color or monochrome. Using the Set Default Colors for Color Monitor command affects both the Session screen and the Data Editor.

When you are finished customizing your screen colors and Data Editor settings and have returned to the MSETUP Main Menu:

- Press Esc to exit MSETUP and return to the DOS prompt, which displays STAT101 as the current directory if you just finished installing STAT101.

Networks

STAT101 is designed to be used by a single user on a single computer. It will not run as a shared application installed on a network server. If more than one user tries to run STAT101 at the same time, STAT101 will terminate.

Making Backup Copies of STAT101

You may make up to two backup copies of STAT101 disk for your personal use, solely for backup or archival purposes, according to the terms explained in your license agreement.

You cannot make backup copies of STAT101 by simply copying the files on your original disk. Instead, you must use the Install Backup command, described in this section.

- Put your STAT101 disk into drive A (or drive B).
- Go to the DOS prompt (C:\STAT101> if you just finished installing STAT101).
- Type **A: INSTALL** or **B: INSTALL**.
- Press [Enter].

If you see Personalization instructions on your screen asking you to enter your name, proceed according to those instructions (see Personalization in the Installation section). You only personalize once, but you must complete it in order to install STAT101 or to make a backup disk.

The STAT101 Installation screen appears.

- Press [Enter] to continue.

The Install Main Menu appears.

- Press [↓] twice to highlight Make a Backup Copy of STAT101 on a Floppy Disk.
- Press [Enter].

The instructions on the screen tell you that you may make only two copies of the STAT101 program disk. The option to continue with the backup process is highlighted, but you also have the opportunity to exit. To continue making the backup copy:

- Press [Enter].
- Select the correct floppy drive by pressing [↑] or [↓] to highlight it.
- Press [Enter].

The backup process begins.

- Follow the instructions on the screen until your backup copy is made.

If you make a mistake, perhaps by inserting a disk without enough space to contain the backup copy, or putting it in the wrong drive, STAT101 walks you through the process of correcting the error. Follow the instructions on the screen carefully.

When STAT101 is finished backing up the program disk, it returns to the INSTALL Main Menu. To exit the INSTALL program and start STAT101:

- Press ⊡ until you have highlighted the last option on the menu, Exit to DOS.
- Press [Enter].

The DOS prompt appears.

Starting and Stopping STAT101

If you have not installed STAT101 on your computer's hard disk, you should install it now. Refer to the Installation section for instructions.

Starting STAT101

To start STAT101:

- If it is not already on, turn on your computer.
- Go to the DOS prompt (C:\>).

Don't worry if a directory name appears after the C:\ part of the prompt. The instructions that follow move you into the correct directory. (See Using DOS for information on directories.)

If you are running Microsoft Windows, double-click the DOS program icon in the Main window of the Program Manager. STAT101 can run under Windows if you have enough memory available (you need 460K of conventional memory). If you do not have enough memory, you may need to exit Windows before running STAT101.

Use the DOS CD command (for Change Directory) to make STAT101 the current directory:

- Type **CD \STAT101** after the DOS prompt and press [Enter].

The prompt now looks like this:

⇨ C:\STAT101>

10 Chapter 1: Getting Started

If you are starting STAT101 right after installing it, the DOS prompt automatically displays STAT101 as the current directory, so using the CD command didn't change anything. Making STAT101 the current directory before starting STAT101 ensures that you can easily access all files in the STAT101 directory.

- Type **STAT101** at the prompt and press [Enter].

STAT101 starts and displays the opening screen, which contains release and license information.

- Press [Enter] to continue.

STAT101 displays the STAT> prompt, shown in Fig. 1–4, which indicates that you can now enter commands.

FIGURE 1–4

```
STAT101   Release 1.1 *** COPYRIGHT (c) - Minitab Inc. 1993
Storage Available:   2000
APRIL 9, 1993
Use the ESCape key to toggle between STAT101 and the Data Editor

STAT>
```

The screen in Fig. 1–4, called the *Session screen* because this is where you do most of the work during a session, displays information about STAT101, including the current release number, the amount of storage available in your worksheet, and the prompt STAT>. STAT101 uses several different prompts; STAT> indicates that you can now begin entering commands. (See Chapter 4, Issuing Commands, for more information on prompts.)

Stopping STAT101

When you are ready to end the session:

- Type **STOP** after the STAT> prompt and press [Enter].

STAT101 asks if you are sure you want to stop. Before you answer, consider whether you have saved all your data and work. If you haven't, you may want to save the current worksheet or any open files before proceeding:

- Press [N] to remain in STAT101 or [Y] to exit, and then press [Enter].

If you answered no, STAT101 returns you to the STAT> prompt, and you can save any work before ending the session. If you answered yes, STAT101 returns you to the DOS prompt or your computer's menu.

The next chapter takes you through two sample STAT101 sessions, giving you the opportunity to get immediate hands-on experience working with data, plots, and statistical analysis techniques.

2

Sample Sessions

How to Use the Sample Sessions

This chapter includes two sample sessions, Basic and Advanced. Both are hands-on introductions to the STAT101 software. Each should take you less than an hour. By following the steps outlined here, you experience a STAT101 session as the data analyst, and start learning your way around. The Basic sample session starts with opening, exploring, modifying, and saving a worksheet. In the Advanced sample session you will try some of STAT101's statistical commands. If you are using this manual as part of a course, you may want to delay going through the Advanced session until you have had some experience with statistics. Be sure to complete the Basic session before proceeding with the Advanced one.

All of the commands used in this session are covered in detail in later chapters of this manual.

You will need the STAT101 worksheet PULSE.MTW that comes with the sample data in every release of STAT101. If, after following the directions in the Basic session, you can't find this file, see your instructor for help. Appendix B contains a description and listing of the data.

To proceed through the sessions, follow the instructions after the bullets (•). They will tell you to type certain commands, shown in **boldface.**

Press [Enter] after every line you type. Although this manual shows all commands in capital letters, you can use any combination of upper- and lowercase letters.

If you make a mistake while typing a command, backspace over it, using the backspace key [←Backspace], and retype it. If you have already pressed [Enter], retype the command on a new command line.

If you want to start the session over, type RESTART after the STAT> prompt, and STAT101 will clear the worksheet. You will have to return to the beginning of the session.

Basic Sample Session

Objectives

The Basic sample session shows you how to

- save and annotate your session work in an outfile
- retrieve and save a worksheet
- look at data graphically with a dotplot
- correct the worksheet by changing and adding data
- getting online help
- obtain basic descriptive statistics

Getting Started Using STAT101

Measuring pulse rates before and after vigorous activity is one way to assess our physical condition. Many things affect what kind of shape we are in, including diet, daily exercise, and habits like smoking or drinking. Students in an introductory statistics course decided to explore their own physical condition in relation to a few of these variables.

The Experiment

Each student in the class recorded his or her height, weight, gender, smoking preference, usual activity level, and resting pulse. Then they participated in a simple experiment: the students all flipped coins, and those whose coins came up heads ran in place for one minute. Then the entire class recorded their pulses once more.

One of the students volunteered to enter the collected data into STAT101. The resulting worksheet contained the following information (C1 means column 1):

Variable	Description
C1 'PULSE1'	Resting pulse rate
C2 'PULSE2'	Second pulse rate
C3 'RAN'	1 = ran in place
	2 = did not run in place
C4 'SMOKES'	1 = smokes regularly
	2 = does not smoke regularly
C5 'SEX'	1 = male
	2 = female
C6 'HEIGHT'	Height in inches
C7 'WEIGHT'	Weight in pounds
C8 'ACTIVITY'	Usual level of physical activity:
	1 = slight
	2 = moderate
	3 = a lot

The volunteer who entered the data gave the worksheet file the name PULSE. Now it's up to you to start looking for trends in the data.

Starting STAT101

Begin by starting STAT101:

- Type **CD \STAT101** after the DOS prompt and press [Enter].

This changes the current directory to STAT101, and from here you can start the program:

- Type **STAT101** after your computer's prompt and press [Enter].

The opening screen appears, giving you general information about STAT101. To continue:

- Press [Enter].

The Session screen opens, as in Fig. 2–1.

FIGURE 2–1

```
STAT101 Release 1.1 *** COPYRIGHT (C) Minitab Inc. 1993
Storage Available: 2000
APRIL 9, 1993

Use the ESCape key to toggle between STAT101 and the Data Editor

STAT>
```

The Session screen displays information about STAT101, followed by the prompt STAT>. This prompt is STAT101's way of telling you that it is ready for you to enter a command.

The first commands you type after this initial prompt typically involve either retrieving a file containing data that has already been saved or entering a new data set that you will work with as long as the session lasts.

Saving Your Session in a File

STAT101 does not automatically save your work unless you tell it to, so it is a good practice to begin by creating a file that stores a record of the entire session. The OUTFILE command tells STAT101 to store a record of everything that follows in whatever file you indicate until you tell it to stop. Always enclose file names in single quotes.

- Type **OUTFILE 'EXERCISE'** and press Enter.

The command you type appears on the Session screen:

⇨   STAT> OUTFILE 'EXERCISE'

(This manual uses an arrow (⇨) in the margin to point to things you will see on the session screen.)

From this point on, STAT101 stores a record of your session in an *outfile*, an ASCII text file named EXERCISE (see Chapter 5, Handling Files and Printing, for information on ASCII text files).

Annotating Your Work

To include your name in the outfile:

- Type **# your name** and press Enter.

The pound sign (#) tells STAT101 that what follows is not a command, but just a comment that you want included in your record of the session.

You can enter as many notes as you want throughout the session if you want to record observations or reminders. If you are turning in your outfile for a class assignment, you could enter your class section and professor's name as a note too.

Retrieving Data from a File

Now retrieve the saved worksheet named PULSE.

- Type **RETRIEVE 'PULSE'** and press Enter.

Be sure to put single quotes around the file name. STAT101 displays basic worksheet information, as shown in Fig. 2–2.

FIGURE 2–2

```
STAT> RETRIEVE  'PULSE'

WORKSHEET SAVED 3/ 3/1993
Worksheet retrieved from file: PULSE.MTW
```

STAT101 adds the file extension MTW for file type identification purposes. See Chapter 5, Handling Files and Printing, for more information on STAT101's file extensions.

If you have trouble retrieving the PULSE worksheet file, the file may be located in a directory other than the STAT101 directory. If this is the case, type the RETRIEVE command as shown, substituting the correct path information. See Chapter 5, Handling Files and Printing, for information on paths.

Viewing Data

To remind yourself of what the data set contains, use the INFO command to see the variables in the current worksheet. INFO lists all column numbers, column names, and the number of rows in each column.

- Type **INFO** and press Enter.

STAT101 displays basic worksheet information on the Session screen, shown in Fig. 2–3 as follows.

FIGURE 2-3

```
STAT> INFO

COLUMN    NAME       COUNT
C1        PULSE1     92
C2        PULSE2     92
C3        RAN        92
C4        SMOKES     92
C5        SEX        92
C6        HEIGHT     92
C7        WEIGHT     92
C8        ACTIVITY   92

CONSTANTS USED: NONE
```

Now look at the data set itself by going to the Data Editor, a spreadsheet that displays your data's rows and columns.

- Press Esc.

If you type a character or a space by mistake after the STAT> prompt before pressing the Esc key, STAT101 produces a "\". Just press Enter to go to a new STAT> prompt, and press Esc again.

The Data Editor appears, showing the PULSE data set, as shown in Fig. 2-4.

FIGURE 2-4

	PULSE1	PULSE2	RAN	SMOKES	SEX	HEIGHT	WEIGHT	ACTIVITY				
	C1	C2	C3	C4	C5	C6	C7	C8	C9	C10	C11	C12
1	64	88	1	2	1	66.00	140	2				
2	58	70	1	2	1	72.00	145	2				
3	62	76	1	1	1	73.50	160	3				
4	66	78	1	1	1	73.00	190	1				
5	64	80	1	2	1	69.00	155	2				
6	74	84	1	2	1	73.00	165	1				
7	84	84	1	2	1	72.00	150	3				
8	68	72	1	2	1	74.00	190	2				
9	62	75	1	2	1	72.00	195	2				
10	76	118	1	2	1	71.00	138	2				
11	90	94	1	1	1	74.00	160	1				
12	80	96	1	2	1	72.00	155	2				
13	92	84	1	1	1	70.00	153	3				
14	68	76	1	2	1	67.00	145	2				
15	60	76	1	2	1	71.00	170	3				
16	62	58	1	2	1	72.00	175	3				
17	66	82	1	1	1	69.00	175	2				
18	70	72	1	1	1	73.00	170	3				
19	68	76	1	1	1	74.00	180	2				
20	72	80	1	2	1	66.00	135	3				
21	70	106	1	2	1	71.00	170	2				
22	74	76	1	2	1	70.00	157	2				

`<F1> for HELP; <F10> for MENU      Last Column: C8   Last Row: 92`

The Data Editor presents a worksheet consisting of vertical columns and horizontal rows. In general, each column contains data for one variable and each row contains the information for one case. The Data Editor provides a convenient way to enter data, name columns, browse through data, and edit individual cells. A *cell* is the intersection of a row and a column; the *active cell* is the highlighted one.

Glancing at the Data Editor, you notice that the person entering the data first listed the males (SEX = 1) who ran in place (RAN = 1).

To see the rest of the data:

- Press and hold down ⬇ for a few seconds. (If you are using the arrows on your keypad, be sure the [Num Lock] key is turned off.)

The Data Editor scrolls through the data, moving the highlighted cell (the active cell) down row by row.

- Press [Pg Dn] to move more quickly through the data, a screen at a time.

As you move through all of the data, notice that the person entering the data then listed females who ran, then males who didn't run, and then females who didn't run.

To return to the top of the Data Editor, use a *key combination,* a shortcut that you execute by pressing two keys at the same time:

- Press and hold down [Ctrl] as you press [Home].

The cursor returns to the beginning of the data set, and the active cell is C1 row 1.

This manual indicates key combinations by a + between the two keys: [Ctrl] + [Home], for example, moves the active cell to C1 row 1. [Ctrl] + [End] moves the active cell to the very end, C1 row 92.

Starting Your Analysis

There are a number of different approaches you could take in analyzing this data set. You decide to start by examining C1 PULSE1, the variable containing the first pulse reading (that is, resting pulse). Is there any variation, perhaps by activity level? Previous medical research suggests that the more active people are, the lower their resting pulses are. Is this true for your class?

Looking at a Dotplot

Many statisticians feel that it is a good practice to begin analysis by looking at the data graphically. You decide to look at the distribution of the students' resting pulses by activity level with a dotplot, using the DOTPLOT command with the subcommand BY. (Subcommands give main commands more information.)

You issue commands from the main STAT101 screen, the Session screen. To return to it:

- Press [Esc].

The STAT> prompt appears, ready for you to enter a command.

You only have to type the first four letters of any STAT101 command (and sometimes fewer, as in COS for COSINE). The rest of this sample session will use that shortcut from time to time. Be sure to type the punctuation correctly.

- Type **DOTP 'PULSE1';** and press [Enter].
- Type **BY 'ACTIVITY'.** and press [Enter].

The semicolon (;) tells STAT101 that a subcommand will follow, and the period (.) says that you are done typing the command. When you type the semicolon (;), the STAT> prompt changes to SUBC>. The prompt STAT> says, "I'm ready for a command," while SUBC> says, "I'll accept subcommands until you tell me to stop by typing a period."

The dotplot appears, as shown in Fig. 2–5.

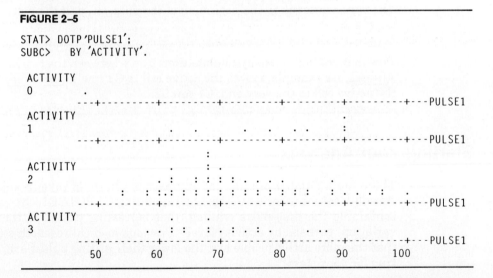

FIGURE 2–5

```
STAT> DOTP 'PULSE1';
SUBC>    BY 'ACTIVITY'.
```

The dotplots don't offer much conclusive evidence, although you can see that most of the students classified their activity level as "moderate," or activity level 2. Recall that level 1 is "slight" and 3 is "a lot" of physical activity. There is a wide range in the resting pulse readings for all three groups.

Correcting the Worksheet

But what is activity level 0, the first dotplot? Looking back at the experiment description, you notice that there is no level 0. Moreover, it corresponds with a resting pulse of 48 in row 54, a rather low reading. The person entering the data could have made a few mistakes—easy enough to do, and something that statisticians must always watch for. The entry of 0 could have a significant effect on your subsequent analysis.

Making Global Changes Using CODE

You decide to delete the 0 activity level and replace it with the missing value symbol, an asterisk (*), since you're not sure what it should be. One way to change the 0 to * is by using the CODE command. Tell STAT101 to replace every 0 it finds in the column ACTIVITY with the symbol *, and put it back in the same column.

- Type **CODE (0) '*' 'ACTIVITY' 'ACTIVITY'** and press Enter.

⇨ STAT> CODE (0)'*' 'ACTIVITY' 'ACTIVITY'

The first time you type ACTIVITY, you are identifying the column containing the values you want to code, and the second time you type it, you are telling STAT101 to place the new values in the same column. STAT101 replaces each 0 with an asterisk; to verify this, use the INFO command once more, getting information only on C8 ACTIVITY:

- Type **INFO C8** and press Enter.

The worksheet information appears on the screen, as shown in Fig. 2–6.

FIGURE 2–6

```
STAT> INFO C8

COLUMN    NAME        COUNT     MISSING
C8        ACTIVITY    92        1
```

STAT101 now reports one missing value in C8. It looks like you've taken care of the problem.

Changing Cell Values

You still need to decide how to handle the surprisingly low resting pulse entry of 48 that corresponded to the 0 activity level. Leafing back through the survey sheets the students turned in, you find the sheet corresponding with this person. That student apparently failed to enter an activity level (which may be why the data entry person entered it as 0). Moreover, the resting pulse should have been 58, not 48.

To fix this mistake, found in row 54 of the data set, return to the Data Editor:

- Press [Esc].

You could scroll down the worksheet, using [↓] or [Pg Dn] to get to row 54, but it may be quicker to use the Go To command.

To use Go To to move to C8 row 54:

- Press [F10].

Five menu command names appear at the bottom of the screen in a single row (Format is the first, and then GoTo, Help, Name, and Reformat), as shown in Fig. 2–7.

FIGURE 2–7

```
     PULSE1 PULSE2 RAN SMOKES SEX HEIGHT WEIGHT ACTIVITY
        C1     C2   C3   C4   C5   C6     C7      C8    C9  C10  C11  C12
  1     64     88    1    2    1  66.00   140      2
  2     58     70    1    2    1  72.00   145      2
  3     62     76    1    1    1  73.50   160      3
  4     66     78    1    1    1  73.00   190      1
  5     64     80    1    2    1  69.00   155      2
  6     74     84    1    2    1  73.00   165      1
  7     84     84    1    2    1  72.00   150      3
  8     68     72    1    2    1  74.00   190      2
  9     62     75    1    2    1  72.00   195      2
 10     76    118    1    2    1  71.00   138      2
 11     90     94    1    1    1  74.00   160      1
 12     80     96    1    2    1  72.00   155      2
 13     92     84    1    1    1  70.00   153      3
 14     68     76    1    2    1  67.00   145      2
 15     60     76    1    2    1  71.00   170      3
 16     62     58    1    2    1  72.00   175      3
 17     66     82    1    1    1  69.00   175      2
 18     70     72    1    1    1  73.00   170      3
 19     68     76    1    1    1  74.00   180      2
 20     72     80    1    2    1  66.00   135      3
Format  GoTo Help Name Reformat
Format a column
Use arrow keys to move; <Enter> or First Letter to select; <ESC> to exit menu.
```

- Press [→] to highlight GoTo and press [Enter].
- Type **1** for column C1 PULSE1 and press [Enter].
- Type **54** for row 54 and press [Enter].

STAT101 moves the active cell to row 54, column C1.

To change the value from 48 to 58:

- Type **58**.

- Press ⌜Enter⌝ to tell STAT101 to accept the new value.

You could repeat the DOTPLOT command if you want to take another look at the data set with the corrections made (if you do, first press ⌜Esc⌝ to return to the Session screen).

Adding Data

Just as you think you have the worksheet cleaned up, your professor informs you that two students who missed class the day of the experiment want their data included. The two students report the following data, shown in Fig. 2–8:

FIGURE 2–8

PULSE1	PULSE2	RAN	SMOKES	SEX	HEIGHT	WEIGHT	ACTIVITY
73	84	1	2	1	65	138	2
72	70	2	2	1	72	185	3

To move to the end of the worksheet:

- Press ⌜Ctrl⌝ + ⌜End⌝.

STAT101 moves you to the lower right of the worksheet.

- Press ⌜Ctrl⌝ + ⌜←⌝ to move to the beginning of row 92.
- Press ⌜↓⌝ to move to the first cell of row 93, which is now the active cell.

Before you enter the new rows of data, notice the small arrow in the upper-left corner of the worksheet, shown in Fig. 2–9. This is the *data-entry arrow,* which controls the direction in which data is entered.

FIGURE 2–9

	PULSE1	PULSE2	RAN	SMOKES	SEX	HEIGHT	WEIGHT	ACTIVITY				
→	C1	C2	C3	C4	C5	C6	C7	C8	C9	C10	C11	C12
72	94	92	2	1	2	62.00	131	2				
73	60	66	2	2	2	62.00	120	2				
74	72	70	2	2	2	63.00	118	2				
75	58	56	2	2	2	67.00	125	2				
76	88	74	2	1	2	65.00	135	2				
77	66	72	2	2	2	66.00	125	2				
78	84	80	2	2	2	65.00	118	1				
79	62	66	2	2	2	65.00	122	3				
80	66	76	2	2	2	65.00	115	2				
81	80	74	2	2	2	64.00	102	2				
82	78	78	2	2	2	67.00	115	2				
83	68	68	2	2	2	69.00	150	2				
84	72	68	2	2	2	68.00	110	2				
85	82	80	2	2	2	63.00	116	1				
86	76	76	2	1	2	62.00	108	3				
87	87	84	2	2	2	63.00	95	3				
88	90	92	2	1	2	64.00	125	1				
89	78	80	2	2	2	68.00	133	1				
90	68	68	2	2	2	62.00	110	2				
91	86	84	2	2	2	67.00	150	3				
92	76	76	2	2	2	61.75	108	2				
93												

Last Column: C8   Last Row: 92

Make the data-entry arrow point to the right if it is not:

- Press [F3] until the arrow points to the right ([F3] toggles the arrow back and forth between right and down).

To enter the two new rows of data:

- Type the first row, one number at a time, pressing [Enter] after each number: **73 84 1 2 1 65 138 2**.
- Press [Ctrl] + [Enter] to move to the first cell of the next row, row 94.
- Type the second row: **72 70 2 2 1 72 185 3** (again be sure to press [Enter] after typing the last value).

If you make a mistake before pressing [Enter] to accept the value, press the [←Backspace] key to delete the incorrect values, and retype the entry. If you find after pressing [Enter] that you made a mistake, use the arrow keys to highlight the cell containing the mistake, type the correct entry, and press [Enter] to accept it.

You have updated your worksheet and are ready to continue.

Saving Data

It is a good idea to save the worksheet in a permanent file whenever you make any changes you intend to keep. So far your changes exist only in your computer's memory and not in a permanent file.

To save the PULSE worksheet as NEWPULSE (so you don't change the contents of the original worksheet), you must first return to the Session screen:

- Press [Esc].
- Type **SAVE 'NEWPULSE'** and press [Enter].

⇨ STAT> SAVE 'NEWPULSE'

Worksheet saved into file: NEWPULSE.MTW

If someone has already gone through this session, there may already be a NEWPULSE in your STAT101 directory, in which case STAT101 asks if you want to replace the existing file. Either press [Y] and then [Enter] for yes to replace it (in which case your NEWPULSE replaces the old NEWPULSE), or [N] and then [Enter] for no, in which case you should repeat the SAVE command with a different file name in single quotes.

You've created a new file, NEWPULSE. The one you originally opened, PULSE, remains unchanged.

Getting Online Help from STAT101

You want to look at some basic statistics for the resting pulse variable, but you're not sure which command to use.

The HELP command can help you decide:

- Type **HELP** and press [Enter].

⇨ STAT> HELP

Follow the instructions on the screen to see how the Help facility works by getting help on Help. You could type HELP HELP, or you could use a shortcut key. The [F3] key copies the previous line onto a new line.

- Press [F3].

The HELP command you just typed appears.

- Type a space.
- Type **HELP** to tell STAT101 what you want help on (help on Help).

⇨ STAT> HELP HELP

- Now press [Enter].

The first HELP tells STAT101 to call up the Help function, and the second directs you to the topic, in this case, help on Help. Using the [F3] key here didn't save all that much time, but for very long commands you may find it useful. See your DOS manual for more information on other shortcut keys like [F3].

The HELP HELP screen lists the various categories available to you. Suppose you want to look at some basic descriptive statistics for the resting pulse variable, but are not sure how to do it. You decide to get help on COMMANDS to find out which command to use. You first need to answer the MORE? prompt.

- Type [N] for no and press [Enter] when STAT101 asks you if you want more information.

The HELP COMMANDS screen lists all the command categories.

- Type **HELP COMMANDS** and press [Enter].

- Press [Y] for yes and press [Enter] when STAT101 asks if you want to see more.

The command list appears, as shown in Fig. 2–10.

FIGURE 2–10

1 General Information	10 Analysis of Variance
2 Entering Data	11 Nonparametrics
3 Outputting Data	12 Tables
4 Editing and Manipulating Data	13 Time Series
	14 Distributions & Random Data
5 Arithmetic	15 Sorting
6 Column and Row Operations	16 Miscellaneous
7 Plots and Histograms	17 Stored Commands and Loops
8 Basic Statistics	18 How Commands are Explained in HELP
9 Regression	

Item 8, Basic Statistics, looks like a good place to start.

- Type **HELP COMMANDS 8** and press [Enter].

The list of basic statistics, shown in Fig. 2–11, appears.

FIGURE 2–11

```
STAT> HELP COMMANDS 8

COMMANDS 8. Basic Statistics

DESCRIBE      (descriptive statistics for each column)
ZINTERVAL     (confidence interval, sigma known)
ZTEST         (separate test on data in each column)
TINTERVAL     (t confidence interval)
TTEST         (separate t-test on data in each column)
TWOSAMPLE     (test and c.i.)
TWOT          (test and c.i.)
CORRELATION   (Pearson correlation between pairs of columns)
```

The information STAT101 displays tells you that DESCRIBE may be the command you need. Now get help on that specific command:

- Type **HELP DESCRIBE** and press [Enter].

⇨ STAT> HELP DESCRIBE

The DESCRIBE Help screen tells you that STAT101 offers an array of basic statistics, and lists them. The DESCRIBE command provides exactly the information you're looking for: summary statistics like the number of values in each column, the number missing, the mean, the median, and so on.

Basic Descriptive Statistics

The DESCRIBE command includes the subcommand BY, which gives you the opportunity to produce separate summary statistics for the resting pulses of students at each activity level. Use the four-letter shortcut to describe resting pulse by activity level:

- Type **DESC 'PULSE1';** and press Enter (be sure to include the semicolon).

- Type **BY 'ACTIVITY'.** and press Enter (be sure to include the period).

The statistics appear in table form, as shown in Fig. 2–12.

FIGURE 2–12

```
STAT> DESC 'PULSE1';
SUBC>    BY 'ACTIVITY'.
```

	ACTIVITY	N	MEAN	MEDIAN	TRMEAN	STDEV	SEMEAN
PULSE1	1	9	79.56	82.00	79.56	10.48	3.49
	2	62	72.74	70.00	72.36	10.89	1.38
	3	22	71.59	71.00	71.25	9.39	2.00
	*	1	58.000	58.000	58.000	*	*

	ACTIVITY	MIN	MAX	Q1	Q3
PULSE1	1	62.00	90.00	70.00	90.00
	2	54.00	100.00	65.50	80.00
	3	58.00	92.00	63.50	76.50
	*	58.000	58.000	*	*

When you examine the output it does not surprise you; it seems to conform to what medical research would predict. Take a look at the column labeled MEAN. Those at activity level 1 (slight physical activity) have the highest mean resting pulse, 79.56 beats per minute, while those at activity level 3 (a lot of activity) have the lowest, 71.59 beats per minute. But subjects who are moderately active (level 2) have a mean resting pulse of 72.74, very near to that of level 3. So while the resting pulse does decrease when activity level rises, is the decrease significant?

The Advanced sample session uses statistical techniques to start answering that question. You can either proceed to it now, or wait until you have learned more about advanced statistical techniques.

For now, end the outfile EXERCISE that you started earlier by using the NOOUTFILE command:

- Type **NOOUTFILE** and press Enter.

STAT101 saves your work in the file EXERCISE.LIS.

To exit STAT101:

- Type **STOP** and press [Enter].

STAT101 asks if you are sure (to give you a chance to save worksheet data in case you forgot).

- Press [Y] and press [Enter].

This ends your first STAT101 session. PULSE.MTW is in its original form, NEWPULSE.MTW contains the updated worksheet, and EXERCISE.LIS contains a record of the session, with any notes you may have added. (STAT101 automatically adds the file extensions.)

Advanced Sample Session

The Advanced sample session involves statistical data analysis techniques. Do not go through this session until you have gone through the Basic sample session, because the data set used here (NEWPULSE) is based on changes made to the PULSE data set that came in your STAT101 package. You may want to quickly review the progress made in the Basic session to remind you of the task at hand.

Objectives

The Advanced sample session shows you how to

- perform a oneway analysis of variance
- plot a column using a boxplot, categorizing the data in groups
- evaluate the difference between two means, using a t-test
- create new columns and column subsets
- create scatter plots and multiple scatter plots
- perform regression analysis

Analysis of Variance

When you finished the Basic sample session, your next step was going to be examining the difference among resting pulse rates of people with different activity levels. Before you proceed with that plan, re-open the outfile EXERCISE to add the work you do in this session to that file:

- Start STAT101 (see the Basic sample session for help).

- Type **OUTFILE 'EXERCISE'** after the STAT> prompt and press [Enter].

STAT101 will append all your commands and output to the end of the file you created in the Basic session. To open the data set you worked with in the Basic session:

- Type **RETR 'NEWPULSE'** and press [Enter].

If you saved the PULSE worksheet after making changes to it under a different name when going through the Basic session, type that name instead of NEWPULSE.

Now proceed with your analysis.

STAT101's ONEWAY command is one method you can use to compare the resting pulses of the three different activity levels.

- Type **ONEW 'PULSE1' 'ACTIVITY'** and press [Enter].

Figure 2–13 shows the analysis of variance table and 95% confidence intervals.

FIGURE 2–13

```
STAT> ONEW 'PULSE1' 'ACTIVITY'

ANALYSIS OF VARIANCE ON PULSE1
SOURCE      DF       SS       MS        F       p
ACTIVITY     2      433      217     1.95    0.148
ERROR       90     9971      111
TOTAL       92    10404

                                        INDIVIDUAL 95 PCT CI'S FOR MEAN
                                        BASED ON POOLED STDEV
LEVEL        N     MEAN     STDEV    ---------+---------+---------+-------
  1          9    79.56    10.48                        (-----------*-----------)
  2         62    72.74    10.89              (---*----)
  3         22    71.59     9.39      (-------*-------)
                                     ---------+---------+---------+-------
POOLED STDEV =    10.53                     72.0      78.0      84.0
```

Because the p-value of .148, reported in the third line of output, is greater than the commonly used α value of .05, and because the confidence intervals overlap, you see no evidence to suggest that there is a significant difference among mean resting pulse rates for different levels of activity.

There could be many reasons for this. When the students originally classified themselves as being very active, moderately active, or only slightly active, they may not have known what criteria to use to judge. If you decide to pursue this subject, you may want to collect more specific

Analysis of Variance

data on physical activity, like average number of hours spent exercising each day, types of exercise, and so on.

Boxplots

You may not have accounted for other possible causes of variability, like gender, so exploring that might be a good next step. Up until now you have been examining the relationship between resting pulse and activity level. Now try looking graphically at resting pulse and gender, using the BOXPLOT command. Do males and females have differing resting pulses?

- Type **BOXP 'PULSE1';** and press Enter.
- Type **BY 'SEX'.** and press Enter.

Figure 2–14 shows the two boxplots that appear.

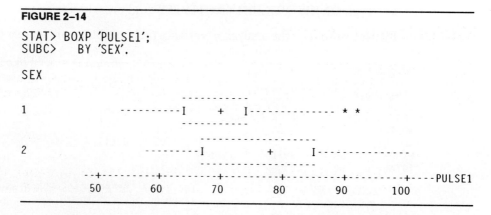

FIGURE 2–14

It appears that males (1) have a lower median resting pulse than females (2).

T-Tests

You can use a two-sample t-test to help you evaluate whether the difference in mean resting pulse rates between genders is statistically significant.

- Type **TWOT 'PULSE1' 'SEX'** and press Enter.

Figure 2–15 shows the two-sample t-test results.

30 Chapter 2: Sample Sessions

FIGURE 2-15

```
STAT> TWOT 'PULSE1' 'SEX'

TWOSAMPLE T FOR PULSE1
SEX    N     MEAN    STDEV   SE MEAN
1     59    70.66    9.47      1.2
2     35    76.9    11.6       2.0

95 PCT CI FOR MU 1 - MU 2: (-10.8, -1.6)

TTEST MU 1 = MU 2 (VS NE): T = -2.67  P = 0.0097  DF = 60
```

The mean resting pulse for males (1) is 70.66 beats per minute, while for females (2) it is 76.9 beats per minute. The p-value of .0097, which is smaller than the commonly used α value of .05, suggests that there may be a significant difference in mean resting pulse rates between males and females. (NE in the output means "not equal.")

Manipulating Columns

Is the difference in pulse rates before and after running also lower in men than in women? You decide to examine the change in the students' first and second pulse readings for those who ran in place. Start by creating columns that will contain data only for the runners.

Creating New Columns with LET

First create a column that stores the difference in all pulses; that is, PULSE2 - PULSE1. Call the column DIFF. STAT101's NAME and LET LETcommands make this easy.

- Type **NAME C9 'DIFF'** and press Enter.
- Type **LET 'DIFF' = 'PULSE2' - 'PULSE1'** and press Enter.

⇨ STAT> NAME C9 'DIFF'
    STAT> LET 'DIFF' = 'PULSE2' - 'PULSE1'

Creating Column Subsets with COPY

Now create two new columns, DIFFRAN and SEXRAN, that contain the pulse differences and genders only for those who ran in place. Use the COPY command to tell STAT101 to find the data only for those who ran and to copy it from DIFF and SEX into DIFFRAN and SEXRAN. (In the RAN column, 1 = ran in place while 2 = did not run in place.)

- Type **NAME C10 'DIFFRAN' C11 'SEXRAN'** and press Enter.

- Type **COPY 'DIFF' 'SEX' 'DIFFRAN' 'SEXRAN';** and press [Enter].
- Type **USE 'RAN' = 1.** and press [Enter].

```
STAT> NAME C10 'DIFFRAN' C11 'SEXRAN'
STAT> COPY 'DIFF' 'SEX' 'DIFFRAN' 'SEXRAN';
SUBC>    USE 'RAN' = 1.
```

Take a look at the columns you just created.

- Type **INFO C9-C11** and press [Enter].

Figure 2–16 shows the INFO table.

FIGURE 2–16

```
STAT> INFO C9-C11

COLUMN   NAME       COUNT
C9       DIFF        94
C10      DIFFRAN     36
C11      SEXRAN      36
```

Out of 94 students, only 36 got heads when they flipped their coins, so only 36 ran in place.

Comparing Levels of a Variable

Now that C10 DIFFRAN contains the pulse differences for those who ran in place, and C11 SEXRAN contains the genders of those students, take a graphical look at how gender might affect the pulse difference. Create a boxplot of DIFFRAN by gender.

- Type **BOXP 'DIFFRAN';** and press [Enter].
- Type **BY 'SEXRAN'.** and press [Enter].

Figure 2–17 shows the boxplots.

FIGURE 2-17

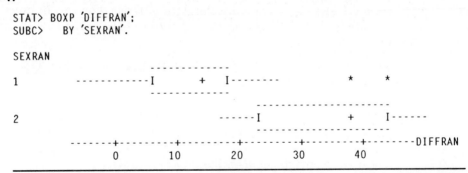

STAT101 displays the boxplot, including two outliers (indicated by asterisks). It appears that the students' pulse increased after having run in place for one minute, but the males (1) did not experience as great an increase as the females (2). Test this difference in mean pulse increase, using a two-sample t-test.

- Type **TWOT 'DIFFRAN' 'SEXRAN'** and press [Enter].

Figure 2-18 shows the two-sample t-test results.

FIGURE 2-18

```
STAT> TWOT 'DIFFRAN' 'SEXRAN'

           TWOSAMPLE T FOR DIFFRAN
SEXRAN    N       MEAN     STDEV    SE MEAN
1         25      12.9     12.2     2.4
2         11      31.9     11.9     3.6

95 PCT CI FOR MU 1 - MU 2: (-28.1, -9.9)

TTEST MU 1 = MU 2 (VS NE): T = -4.38   P = 0.0003   DF = 19
```

The mean pulse increase for males after running in place for one minute is 12.9 beats per minute, while for females it is 31.9. The p-value of .0003 suggests that this difference is significant.

Scatter Plots

Might you be able to predict a runner's post-running pulse (C2 PULSE2) based on his or her resting pulse rate (C1 PULSE1)? To test this, you'll create two new columns, P1-RAN and P2-RAN, that contain the resting pulse and post-running pulses for just the runners. First, name C12 and

C13 with these new column names, and then copy the pulse information from C1 and C2, using only the data of those who ran in place for one minute (coded by RAN = 1). (You can refer to columns by either number or name.)

- Type **NAME C12 'P1-RAN' C13 'P2-RAN'** and press Enter.
- Type **COPY C1 C2 C12 C13;** and press Enter.
- Type **USE 'RAN' = 1.** and press Enter.

Take a look at a scatter plot of the runners' post-running pulse readings (P2-RAN) vs. their resting pulse readings (P1-RAN) to get a feel for the linear relationship between these two variables.

- Type **PLOT 'P2-RAN' 'P1-RAN'** and press Enter.

Figure 2–19 shows the scatter plot as follows.

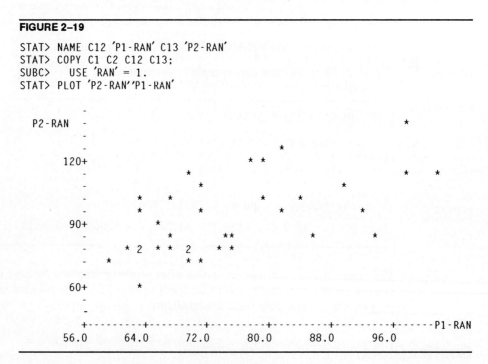

FIGURE 2–19

It does seem that, on the average, post-running pulse rate increases linearly as resting pulse rate increases.

Regression

STAT101's regression capability can help you predict post-running pulse from resting pulse readings. So that you can check your analysis, store the standardized residuals in C14 and the fits in C15. First name the columns.

- Type **NAME C14 'RESIDS' C15 'FITS'** and press Enter.

⇨ STAT> NAME C14 'RESIDS' C15 'FITS'

Now tell STAT101 to regress P2-RAN on one predictor variable, P1-RAN (typing C14 and C15 after the command tells STAT101 where to store the residuals and fits, respectively):

- Type **REGRESS 'P2-RAN' 1 'P1-RAN' C14 C15** and press Enter.

Figure 2–20 shows the regression equation and the analysis of variance table.

FIGURE 2–20

```
STAT> REGRESS 'P2-RAN' 1 'P1-RAN' C14 C15

The regression equation is
P2-RAN = 18.2 + 1.01 P1-RAN

Predictor       Coef        Stdev      t-ratio          p
Constant       18.17        16.86         1.08      0.289
P1-RAN        1.0071       0.2266         4.44      0.000

s = 15.11      R-sq = 36.8%      R-sq(adj) = 34.9%

Analysis of Variance

SOURCE        DF           SS           MS           F          p
Regression     1       4509.8       4509.8       19.76      0.000
Error         34       7761.4        228.3
Total         35      12271.2

Unusual Observations
Obs.  P1-RAN    P2-RAN        Fit  Stdev.Fit   Residual   St.Resid
 29      100    115.00     118.88       6.49      -3.88      -0.28 X

X denotes an obs. whose X value gives it large influence.
```

The regression equation, P2-RAN = 18.2 + 1.01 P1-RAN, tells you that to predict (approximately) the pulse after running for students similar to those in the sample, add 18.2 to the resting pulse (because 1.01 times P1-RAN is essentially the resting pulse). The p-value of 0.000 supports your suspicion that there is a significant relationship between resting pulse and post-running pulse.

Before going further, it would be wise to save your worksheet once again.

- Type **SAVE 'NEWPULSE'** and press Enter.

⇨ STAT> SAVE 'NEWPULSE'

When STAT101 asks if you want to replace the existing file:

- Press Y for yes and press Enter.

⇨ Worksheet saved into file: NEWPULSE.MTW

This replaces the NEWPULSE file created earlier with the new version of the worksheet.

Multiple Scatter Plots

It can be instructive to plot the fitted values along with the actual data.

- Type **MPLOT 'P2-RAN' 'P1-RAN' 'FITS' 'P1-RAN'** and press Enter.

Figure 2–21 shows the MPLOT.

FIGURE 2–21

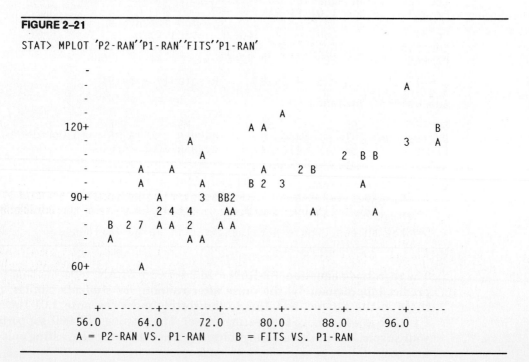

Chapter 2: Sample Sessions

To see the fitted regression line, you could manually draw a line roughly connecting the B's. You might want to look at a residual plot, but for now, look back at the regression results. The R-squared value of 36.8% indicates that this model explains less than 40% of the total variability. Since you learned earlier that men and women have different pulse rates, you suspect that perhaps you should have included gender in the model. You decide to investigate this later. For now, plot post-running pulse vs. resting pulse, marking the data points by gender.

- Type **LPLOT 'P2-RAN' 'P1-RAN' 'SEXRAN'** and press Enter.

Figure 2–22 shows the LPLOT.

FIGURE 2–22

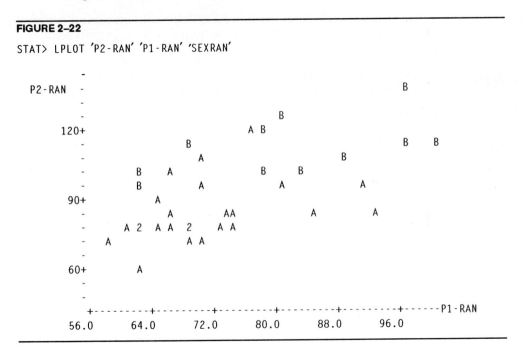

Male pulse rates, marked by A's, appear to be lower than the women's, so, as you expected, gender does seem to account for some of the additional variability.

This is a good time to take a break. Remember that you began the session by opening the EXERCISE outfile. STAT101 has been saving all the steps you took in your analysis in this file, appending it to the end of the file you created in the Basic session.

To stop adding to the EXERCISE file:

- Type **NOOUTFILE** and press Enter.

Multiple Scatter Plots 37

There are many directions you could go from here with this data set. You have not even begun to consider the smoking data, or the weights and heights of the students. But you have seen how a typical STAT101 session progresses, and how to use some basic commands. Try some of what you've learned on the rest of the data set; you may uncover some surprises!

When you are finished:

- Type **STOP**.

STAT101 asks if you are sure (to give you a chance to save any worksheet data).

- Press [Y] and press [Enter].

This ends your STAT101 session. EXERCISE.LIS contains a record of the entire session, appended to the work from the Basic sample session.

You may want to include this output in a report, using a word processing application or text editor. EXERCISE.LIS is a text file that you can print on almost any printer, and import into editing software to work with further.

Now What?

This sample session has given you practice moving around STAT101. Chapters 3, 4, and 5 walk you through specific task-oriented examples that teach you how to use STAT101's worksheet and command structure, the basics of data entry, and how to create and save worksheets and other files. From there, you can browse through Chapter 6, Command Reference, as the need arises, to study additional individual commands.

Because the examples that accompany most of the commands in the manual use data sets included on your program disk and installed with STAT101, you can study and duplicate them while running STAT101 on your own computer. The best way to learn is by trying it yourself.

3

Using the Data Editor

Objectives

This chapter explains how to enter and edit a small data set, using the Data Editor. You will learn how to

- open the Data Editor to view the current worksheet
- navigate around the worksheet
- get help using the Data editor
- enter and edit data in the Data Editor
- use store constants
- work with data formats, including alphabetic and numeric data

The end of the chapter includes a complete keyboard summary.

Opening the Data Editor

When you first start STAT101, it creates a temporary storage area in your computer's memory called the *worksheet,* which stores the data you will work with during your session. The worksheet, like any spreadsheet, is arranged into *columns* (variables) and *rows* (individual cases). A worksheet can also contain *constants* that store single numbers.

There are a number of ways to enter data into the worksheet. The two most common are retrieving it from a STAT101 saved worksheet file or typing it in from the keyboard. This chapter explains how to enter a small data set into the worksheet from the keyboard, using STAT101's Data Editor.

- Start STAT101 (see Starting and Stopping STAT101 in Chapter 1).

The opening screen appears.

- Press [Enter] to open the Session screen.

The STAT101 Session screen appears, as shown in Fig. 3–1.

FIGURE 3–1

```
STAT101   Release 1.1 *** COPYRIGHT (c) - Minitab Inc. 1993
Storage Available:   2000
APRIL 9, 1993

Use the ESCape key to toggle between STAT101 and the Data Editor
STAT>
```

All commands are issued from the Session screen: you type them after the STAT> prompt (see Chapter 4, Issuing Commands).

If you are starting this chapter directly from the Sample session, clear the screen using the RESTART command before you proceed:

- Type **RESTART** and press [Enter].

This clears the worksheet and the screen so that it looks like Fig. 3–1.

Notice the instructions just above the STAT> prompt that tell you, "Use the ESCape key to toggle between STAT101 and the Data Editor."

To open the Data Editor:

- Press the [Esc] key.

If you type a character or a space by mistake after the STAT> prompt before pressing the [Esc] key, STAT101 produces a "\". Just press [Enter] to go to a new STAT> prompt, and press [Esc] again.

The Data Editor opens, showing as many columns and rows as your monitor can display, as shown in Fig. 3–2.

The intersection of a row and a column is called a *cell*. Each cell is identified by its column (C1, C2, C3...) and its row (1, 2, 3...).

Each column contains data for one variable. You can use up to 100 columns and 100 constants with a total of 2000 data points. Each row contains one observation, or case.

The highlighted cell that appears in the upper-left corner of the worksheet when you first open the Data Editor corresponds to column 1, row 1.

FIGURE 3-2

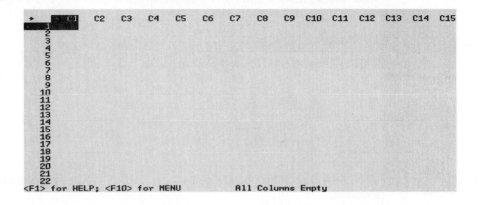

Because this cell is highlighted (or colored or boxed, depending on your monitor type), you can tell it is the *active cell*. Any information you type from the keyboard goes into the active cell. There can be only one active cell at a time.

On some monochrome monitors, the active cell may be marked differently, with the active row and column underlined.

Moving Around

You can easily change the active cell when you want to enter data into a different area of the Data Editor.

- Press →. (If you are using the arrows on your numeric keypad, be sure the Num Lock key is turned off.)

This action highlights C2 row 1. Now move the active cell down:

- Press ↓ and hold it down until you reach row 10.

The active cell is now C2 row 10. You can only see as many columns and rows as your monitor can display. The Data Editor actually has 100 columns. To see more of them:

- Press → and hold it down for a few seconds and then release the key.

The window scrolls horizontally to reveal additional columns; it continues until you release the key. To scroll down the Data Editor:

- Press ↓ and hold it down for a few seconds and then release the key.

Now the window scrolls down to show additional rows. To return to the beginning of the worksheet:

- Press Ctrl + Home.

This returns the active cell to C1 row 1.

You can move directly to any cell, using the GoTo command from the *Data Editor menu,* a collection of five commands.

To open the Data Editor menu:

- Press F10.

The Data Editor menu appears at the bottom of the screen, as shown in Fig. 3–3.

FIGURE 3–3

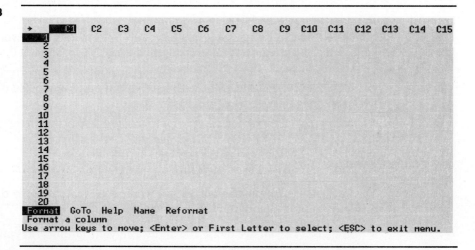

The Data Editor menu opens at the bottom of the screen, with Format, the first command, highlighted, with a description of its use (Format a column) below. To select the GoTo command:

- Press G for GoTo (or press → to highlight GoTo) and then press Enter.

STAT101 prompts you to enter a column number and then a row number. To go to C15 row 3:

- Type **15** and press Enter.
- Type **3** and press Enter.

The active cell is now C15 row 3. There is also a keyboard shortcut for the GoTo command:

- Press [F5].

This opens the GoTo prompt. To go to C40 row 76:

- Type **40** and press [Enter].
- Type **76** and press [Enter].

The active cell is now C40 row 76.

Getting Help

There are other keyboard methods of moving around the Data Editor; the table at the end of this chapter summarizes them. You can also learn more about Data Editor techniques by using the Data Editor's *online help*, information available to you on the screen while you are running the program. (The Session screen also has online help; see Chapter 4, Issuing Commands.)

To open the Data Editor's online help:

- Go to the Data Editor if you aren't already there.
- Press [F1] to open the menu.

The Data Editor Topics menu opens, with the first topic, Introduction, highlighted, as shown in Fig. 3-4.

FIGURE 3-4

```
┌─────Data Editor Topics─────┐
│ Introduction               │
│ Data Editor Screen         │
│ Entering New Data          │
│ Editing Existing Data      │
│ Moving the Cursor          │
│ Data Types                 │
│ Formatting Data            │
│ Quick Reference List       │
│ Troubleshooting            │
└─────────<ESC> to exit──────┘
```

To select a help topic:

- Press [↑] or [↓] to highlight the topic you want.
- Press [Enter] to open that topic.

The Help screen opens, showing a summary of the contents for the topic you chose.

- Press [Pg Dn] to scroll through the contents.
- Press [Esc] at any time to return to the Data Editor Topics menu. (You also return to the menu when you press [Pg Dn] after the last topic.)

To return to the Data Editor from the Topics menu:

- Press [Esc].

Entering Data

When you start STAT101, the current worksheet is empty. To enter data into the worksheet, using the Data Editor:

- Go to the Data Editor, if you are not already there, by pressing [Esc] from the Session screen.
- Press [Ctrl] + [Home] to make C1 row 1 the active cell if it is not already.

Figure 3–5 shows a typical data set of a small company's employee history: ID number, current salary, and years employed with the company.

FIGURE 3–5

ID	Salary	YearsEmp
15	34500	4
44	28000	1
11	32000	1
13	47000	16

How you enter this data into the Data Editor is partly determined by the *data-entry arrow*, a small arrow in the upper-left corner of the Data Editor, shown in Fig. 3–6.

FIGURE 3–6

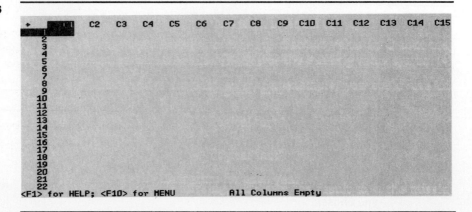

44 Chapter 3: Using the Data Editor

This arrow tells STAT101 whether to accept data by row or by column: when the arrow points to the right, pressing [Enter] moves the active cell to the right; when it points down, pressing [Enter] moves the active cell down.

To change the direction of the data-entry arrow:

- Press [F3].

The arrow changes direction. To enter the employee data row-wise:

- Press [F3] until the data-entry arrow points to the right.

To enter the first row:

- Type **15**.
- If you make a mistake, press [←Backspace] to erase the error, and then retype the number.
- Press [Enter] when the entry is correct.

Pressing [Enter] tells STAT101 to accept the data entry and move the active cell to the right. Had the data-entry arrow been pointing down, the active cell would have moved down.

- Type **34500** and then press [Enter].
- Type **4** and press [Enter].

Now you have entered the first row. To go to the beginning of the next row:

- Press [F6].

This accepts the entry of 4 in C3 row 1 and moves you to the beginning of row 2. If you had been entering data column-wise, pressing [F6] would have taken you to the beginning of column 2.

- Enter the rest of the data in Fig. 3–5, pressing [Enter] after typing each individual data point, and [F6] after the end of each row.

You can press [Ctrl] + [Enter] at the end of a row too; it has the same function as [F6]. Be sure to press [Enter] after typing the last value, so STAT101 accepts it. You can also accept values by pressing the arrow keys to change the active cell. The table at the end of this chapter summarizes key combinations and functions.

You can also enter data in the Session screen by using the READ, SET, and INSERT commands. (See Chapter 6, Command Reference, for more information.)

Editing the Worksheet Sheet

Check over your data to make sure you entered it correctly. To fix a mistake:

- Highlight the cell containing the mistake by using the arrow keys or the GoTo command.
- Retype the entry and press [Enter], and STAT101 accepts the new value.

STAT101's Data Editor uses different modes for different tasks. When you correct an error by activating the cell and typing over the contents, you are in **entry mode,** and what you type replaces existing data in the cell. You can switch to **edit mode** so that what you type doesn't remove the contents of the cell but is added to it.

To edit a cell in edit mode:

- Activate the cell. In this chapter's example, go to C2 row 3, which currently contains 32000.
- Press [F2] to switch to edit mode.

Figure 3–7 shows the cell corresponding to C2 row 3 in edit mode.

FIGURE 3–7

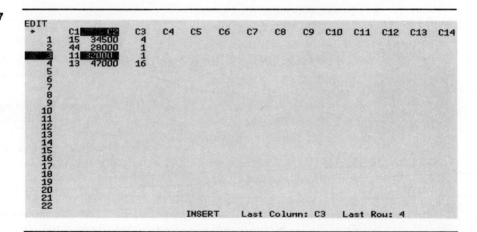

Notice that the word "EDIT" appears in the upper-left corner of the Data Editor, and the word "INSERT" appears on the bottom of the screen. The *cursor,* the small blinking line that appears in the active cell, moves under the first character in the cell. In edit mode, you can either insert characters (in which case all other characters move to the right) or you can overwrite them (in which case the new characters replace the old); the [Ins] key controls whether you are in insert or overwrite mode.

To insert a 1 before 32000 in the example:

- Type **1**.

This changes the number from 32000 to 132000. To change it back:

- Press the [←Backspace] key.

This deletes the 1. Now switch to overwrite mode so that you can change 32000 to 32500.

- Press the [Ins] key to switch to overwrite mode; the INSERT at the bottom of the screen disappears.
- Press [→] to move the cursor under the first 0 in 32000.
- Type **5**.

The 5 replaces the 0 and the new entry says 32500.

- Press [Enter] to accept the new value and return to entry mode. You can also return to entry mode by pressing [F2].

The table at the end of this chapter summarizes other error correction techniques.

- Press [Ctrl] + [Home] to return to the beginning of the worksheet.

To insert or delete entire rows or columns, see the Command Reference entries on ERASE (to delete columns or constants), DELETE (to delete rows from specified columns), and INSERT (to insert new rows in existing columns).

Naming Columns

The row above the column label row in the Data Editor is blank unless you use it to record column names. Names help you remember what variables the individual columns contain.

For example, to give C1 a name:

- Press [Ctrl] + [Home] to move to the top of the worksheet.
- Press [Alt] + [N] to highlight the cell above the C1 label.

You can also press [↑] to move to the name row, or you can select the Name command from the Data Editor menu (press [F10] and then [N] for Name).

If you are following the example in this chapter, C1 presently contains ID numbers for a small company's six employees.

- Type **ID**.

To name the rest of the columns in your worksheet:

- Press [Enter] to move to the next column name cell.
- Type a name for each column, pressing [Enter] after each entry. In the example, name C2 **SALARY** and C3 **YEARSEMP**. Don't forget to press [Enter] after typing the last name.

To return to the data cells:

- Press [Esc] or [↓].

You can also name columns by using the Session command NAME. See Chapter 6, Command Reference, for more information.

The small employer data set that this chapter has been using as an example should now look like Fig. 3–8.

FIGURE 3–8

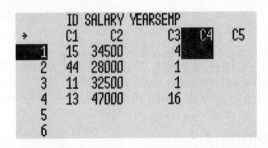

Chapter 5, Handling Files and Printing, discusses saving data sets you have entered, but there is no need to save this data set, as it was only used to illustrate Data Editor techniques.

To return to the Session screen from the Data Editor:

- Press [Esc].

You can also move from the Session screen to the Data Editor by using [Alt] + [D], and from the Data Editor to the Session screen by using [Alt] + [M].

The rest of this chapter discusses two topics related to the worksheet that you can study when you need to know more about them: stored constants and data formats (for using different kinds of numeric data or alphabetic data). The chapter concludes with the keyboard summary.

Stored Constants

Although the Data Editor doesn't display stored constants, the worksheet can store up to 100 of them, each of which can contain a single number. Refer to stored constants as K1, K2, K3, ..., K100. You can assign and view stored constants in the Session screen, using the LET and PRINT commands (see LET and PRINT in Chapter 6, Command Reference). You cannot assign names to stored constants.

STAT101 automatically assigns the values of the missing value code *, e, and π to the last three stored constants: K98 = *, K99 = 2.71828, and K100 = 3.14159. You can assign other values to these constants if you wish, also using the LET command.

Data Formats

STAT101 handles two types of data: numeric (numbers) and alpha (other characters). You will work mostly with numbers, as most data analysis commands handle only numbers. STAT101 assigns a column with the proper format based on the data you first enter. If your first entry contains any non-numeric characters, STAT101 assigns the column an alpha format, and the column width automatically increases to display the longest data item.

The Data Editor automatically adjusts column formats as data are entered, but column formats are not readjusted if data are deleted. The "Reformat" option on the menu resets the column to the default format.

Remember that [F10] opens the Data Editor menu. Make menu selections by pressing the first letter of the menu command or by using the arrow keys to highlight the command you want and then pressing [Enter].

You can override automatic formatting and define your own for any column, using the Format menu command. See Formatting Data in the Data Editor's online help for more information. You cannot, however, change a column from alpha to numeric or vice versa with the Format menu command. To do this, see the Reformatting Alpha Columns section later in this chapter, or use the CONVERT command, described in Chapter 6, Command Reference.

Alpha Data

One row of a column can contain up to 80 alpha characters. The maximum display width of a column is 74 characters, so if a large alpha field does not fit on the screen, STAT101 displays a "+" sign in the right-most position, indicating that there are additional characters that can't be displayed.

You may use any characters (letters, numbers, punctuation symbols, blanks). No column can contain both alpha and numeric data: STAT101 treats numbers appearing in an alpha column (as in a street address) as alpha characters. You can learn more about using alpha data by referring to the Alpha Data information in HELP OVERVIEW 11 in the Session screen.

Numeric Data

When your first entry in a column is a number, STAT101 gives it one of three formats: *integer* (whole numbers), *floating point* (numbers with decimal points), or *exponential* (numbers with base e), depending on the data you enter. Numbers entered in exponential format are displayed in floating point format unless they are larger than 1.0E+12. To display smaller numbers in exponential format, you must specify a fixed exponential format, using the Format command on the menu.

Leading "+" and "–" signs are accepted in numeric columns, but commas are not accepted. Typing 1,000,000 will produce an error. If you entered 1, 2, 4, and 5.5 into a column, STAT101 would initially format the column values as integers, but would then change to floating point format. Remember that the Data Editor uses one format for the entire column.

Reformatting Alpha Columns

Once you have entered an alpha character into a column, STAT101 defines that column as alpha, even if you delete that value later. A column may appear to hold only numbers, yet still be defined as alpha. If you are not sure about a column's formatting status, type the command INFO to see how the column is defined; an "A" appears beside alpha columns.

The easiest way to change a column from alpha to numeric in this case is to WRITE the column to a file, ERASE the column in the worksheet, and then READ the column from the file back into the worksheet. For example, suppose you want to change C1 from alpha to numeric. Create a temporary file, TEMP, using these commands:

- Type **WRITE 'TEMP' C1** after the STAT> prompt and press [Enter].
- Type **ERASE C1** and press [Enter].
- Type **READ 'TEMP' C1** and press [Enter].

The file TEMP holds your data temporarily while you erase the column and its formatting. When you use the READ command to put the data back in the column, STAT101 automatically assigns it a numeric format and ignores all commas.

Data Editor Keyboard Summary

Moving Around

[Alt] + [D] or [Esc]	moves from Session screen to Data Editor
[Alt] + [M] or [Esc]	moves from Data Editor to Session screen
[↑]	accepts new data and moves up one cell
[↓]	accepts new data and moves down one cell
[←]	accepts new data and moves left one cell
[→]	accepts new data and moves right one cell
[Enter]	accepts new data and moves one cell in the direction indicated by data entry arrow, right or down
[F6] or [Ctrl] + [Enter]	accepts new data and moves to beginning of next row or column depending on the direction of data entry arrow
[Pg Up]	moves up one screen
[Pg Dn]	moves down one screen
[Ctrl] + [←]	moves left one screen
[Ctrl] + [→]	moves right one screen
[Home]	moves to upper left of screen
[End]	moves to lower right of screen
[Ctrl] + [Home]	moves to upper left of worksheet
[Ctrl] + [End]	moves to lower right of worksheet

(continued)

`F5`	opens GoTo command to move to specified cell; enter column number and row number and active cell shifts to that location
`Alt` + `N`	moves to column name row

Other Function Keys

`F1`	DataEditor Help
`F3`	changes direction of data-entry arrow
`F4`	removes empty columns from display. Word "COMPRESS" appears at bottom of screen. Press `F4` again to display all columns.
`F7`	inserts empty data cell at active cell and moves remaining values in column down
`F8`	deletes active cell and shifts remaining values in column up; in a name cell deletes the name
`F9`	pens the Format bar
`Shift` + `F7`	inserts a row above the current row and moves remaining rows down
`Shift` + `F8`	deletes a row and moves remaining rows up

Editing Keys

`F2`	switches between entry mode and edit mode
`←`	moves cursor one character left
`→`	moves cursor one character right
`Home`	moves to beginning of cell
`End`	moves to end of cell
`←Backspace`	deletes character to the left of cursor and moves to that position
`Del`	deletes character to the right of cursor
`Ins`	switches between inserting and overwriting characters

4

Issuing Commands

Objectives

This chapter gets you started using STAT101 commands. There are about 150 commands for entering data into a worksheet, manipulating and transforming data, producing graphcal and numeric summaries, and performing a wide range of statistical analyses. This chapter shows you how to

- use STAT101's command language
- issue comands and subcommands
- get Session screen help
- use STAT101 symbols and prompts

Commands and Arguments

STAT101 carries out its procedures when you type any of about 150 commands in its special command language. Most commands are simple words that are easy to remember, like PLOT, TALLY, or PRINT.

Usually you will type a command that operates on one or more *arguments* that you specify. Arguments can be columns, constants, numbers, file names, or text strings. For example, the following command tells STAT101 to retrieve the PULSE data set (used in the sample session):

⇨ `STAT> RETRIEVE 'PULSE'`

The argument in this command is the file name, PULSE, set in single quotation marks.

To try this command:

- Start STAT101 (see "Starting and Stopping STAT101" in Chapter 1) if you haven't already.
- Go to the Session screen if you haven't already (press [Esc] from the Data Editor).
- Type **RESTART** to clear the worksheet if you had any previous data in it.
- Type **RETRIEVE 'PULSE'** after the STAT> prompt.
- Press [Enter].

Pressing [Enter] tells STAT101 to place a copy of the data stored in the PULSE file in the current worksheet.

The RETRIEVE command overwrites any data the worksheet may have contained, so if the current worksheet holds valuable data, you should save it first. See Chapter 5, Handling Files and Printing.

To see the PULSE data set, go to the Data Editor:

- Press [Esc].

When you open a data set, you may see a brief message telling you that STAT101 is reformatting some of the columns to fit the data on the Data Editor.

After reviewing the data, return to the Session screen from the Data Editor:

- Press [Esc].

Learning STAT101's Command Syntax

Most computer programs require that you learn a special language to communicate your instructions. The set of rules by which the language operates is called the *command syntax*. A few commands allow fairly complicated expressions, but most use just a simple list of arguments.

Chapter 6, Command Reference, lists all of STAT101's commands along with the correct syntax for including arguments. For example, the histogram command and its syntax look like this:

HISTOGRAM C...C

The C...C means you can type one or more columns after typing HISTOGRAM.

Try producing a histogram of the data contained in C1, the resting pulse rates of a group of college students:

- Type **HISTOGRAM C1** after the STAT> prompt and press Enter.

STAT101 produces a histogram of the students' resting pulse rates, as shown in Fig. 4–1.

FIGURE 4–1

```
STAT> HISTOGRAM C1

Histogram of PULSE1    N = 92

Midpoint    Count
   50         1   *
   55         2   **
   60        17   *****************
   65         9   *********
   70        23   ***********************
   75        10   **********
   80        11   ***********
   85         6   ******
   90         9   *********
   95         3   ***
  100         1   *
```

The argument in this command was simply C1 (you may also type 'PULSE1' in place of C1, but when you use column names you must put them in quotes).

STAT101 recognizes only the first four letters of any command, so once you get used to the command language you can use four-letter shortcuts, like typing HIST C1 rather than HISTOGRAM C1.

Using Subcommands

Many commands use subcommands that provide STAT101 with more information. To signal STAT101 that you want to use a subcommand:

- End the main command line with a semicolon (;) and then press Enter.

STAT101 then prompts you with the subcommand prompt, SUBC>.

- Type your subcommands, ending each line with a semicolon and then pressing Enter.
- End the last subcommand with a period (.) to tell STAT101 you are finished with the command.

STAT101 then executes the entire command and any subcommands. When typing a command with no subcommand following, you do not have to type any punctuation mark after the command line.

This manual lists subcommand syntax immediately after the main command syntax, slightly indented. The HISTOGRAM command, for example, has several subcommands, one of which is BY.

Try using this command and subcommand on the PULSE data set used in the sample session. To plot a histogram of the resting pulse data (contained in C1) by gender (contained in C5), use the following commands, and be sure to include the punctuation marks indicated:

⇨
```
STAT> HIST C1;
SUBC>   BY C5.
```

If you forget to end the last subcommand with a period, you will receive an error message. Type the period all by itself on the next SUBC> line.

STAT101 first displays the histogram for SEX = 1 (male) and then for SEX = 2 (female). If your monitor can't display both histograms, STAT101 asks if you want to continue the display.

- Press [Y] and press [Enter].

If you want to cancel the whole command, type ABORT as the next subcommand, as shown in Fig. 4–2.

FIGURE 4–2
```
STAT> HIST C1;
SUBC>   ABORT
```

See Chapter 6, Command Reference, for a full list of commands and subcommands.

Typographical Conventions for Commands

This manual uses the following conventions for worksheet and command elements, including arguments:

C   denotes a column, such as C12 or Height

K   denotes a constant, such as 8.3 or K14

E	denotes either a constant or a column
[ ]	encloses an optional argument
CAPITALS	denotes STAT101 command or subcommand, such as SAVE

Syntax Conventions

There are a few conventions that you should know when issuing commands:

- You need only type the first four letters of a command or subcommand. (You can type the full command if you want.)
- Commands are not case-sensitive, so you can type them in lowercase, uppercase, or a combination of both.
- Start each command or subcommand on a new line and press [Enter] when you are done with each line.
- You can continue a command onto the next line with the continuation symbol & (see Symbols later in this chapter).
- Abbreviate groups of consecutive columns or stored constants with a dash. For example, HISTOGRAM C2-C5 is equivalent to HISTOGRAM C2 C3 C4 C5.
- Enclose file names and column (variable) names in single quotation marks (for example, HISTOGRAM 'PULSE1').
- You can add your own notes to any command line (except the LET command), as Fig. 4-3 illustrates; you can also use the # symbol (see Symbols).

STAT101 has a flexible command structure; you can issue commands several different ways. For example, in the PULSE worksheet, columns C1 through C4 are named PULSE1, PULSE2, RAN, and SMOKES. The commands shown in Fig. 4-3 are all equivalent; they each produce histograms of those four columns.

FIGURE 4-3

```
STAT> HISTOGRAM C1 C2 'RAN' 'SMOKES'
STAT> Histogram C1-C4
STAT> HISTOGRAM c1-c4 June 15,1993
STAT> HIST C1-C4
STAT> hist c1-c4
STAT> HISTOGRAM of four variables including 'PULSE1', 'PULSE2' &
CONT> 'RAN' and 'SMOKES'
```

Using Help

STAT101 offers online help at any point in a session, either from the Data Editor (see Chapter 3, Using the Data Editor) or from the Session screen. There are several general commands that you can type in the Session screen for a more complete introduction to STAT101's Help function:

Task	What to type
To learn how to use Help:	**HELP HELP**
For an overview of how to use STAT101:	**HELP OVERVIEW**
For a complete list of all command topics:	**HELP COMMANDS**

For help on specific commands, type **HELP** after a STAT101 prompt and continue with the name of the command or subcommand you want to see explained. For example, for help using the BY subcommand of the HISTOGRAM command, use the command:

⇨ STAT> HELP HISTOGRAM BY

STAT101 then displays information on your screen about the BY subcommand, including its syntax. The Basic sample session in Chapter 2 shows getting help in detail.

Symbols

* Missing Value Symbol

STAT101 uses an asterisk (*) in numeric columns and a blank in alpha columns to represent missing values in a column. Most commands exclude from analysis all rows with a missing value and display the number of excluded points. Unless otherwise noted in this manual, when an arithmetic command operates on a missing value, STAT101 sets the result to *.

& Continuation Symbol

Type the continuation symbol & at the end of any line to indicate that the command continues on to the next line. STAT101 returns with the CONT> prompt (see STAT101 Prompts at the end of this chapter).

Comment Symbol

STAT101 ignores everything you type between the comment symbol # and the end of a line. You can use the symbol on any line. It is particularly useful for adding notes to your session to explain to someone else or to help you remember ideas about your analysis.

STAT101 Prompts

STAT101 uses several different prompts that help you know what kind of input STAT101 expects.

STAT>	Waiting for a command.
SUBC>	Waiting for a subcommand. To return to the STAT> prompt, type either the subcommand with a period (.) after it, or just a period if you decide not to use a subcommand.
DATA>	Waiting for data. To finish entering data and return to the STAT> prompt, type END and press [Enter].
CONT>	Waiting for the rest of the command or data line continued from the previous line. If the previous line ends with the continuation symbol &, STAT101 displays CONT> on the next line.
Continue?	Waiting for [Enter] or Y and then [Enter] to continue displaying output, or N and then [Enter] to discontinue displaying output. Your response to this prompt does not affect the execution of a command; it affects only the display of output on the screen. To suppress this prompt, type OH 0 (output height = 0). The default is OH 24.
More?	Waiting for [Enter] or Y and then [Enter] to continue displaying online help information, or N and then [Enter] to discontinue displaying help.

5

Handling Files and Printing

Objectives

Whenever you are dealing with a data set that you plan to use in the future, you should periodically save it in a more permanent form, either printed, or as a file on a disk. The current worksheet in any STAT101 session exists only in your computer's temporary memory, and would be lost if there were a power failure. STAT101 can store worksheet data in several different file formats, and you can also store the steps you took in analyzing data. This chapter shows you how to

- save and retrieve STAT101 worksheet files
- import and export data, using ASCII text files
- combine data from different worksheets
- save your STAT101 session to a file
- delete files, using DOS
- print data, commands, and STAT101 output

File Types

There are a number of file types available to you, depending on the nature of your work and what you need to save (the data itself, the command sequence from a given session, STAT101 output, and so on). The table at the end of this chapter summarizes the five STAT101 file types.

For each file type, STAT101 adds a *file extension* to the name, a three-character suffix to the file name that helps categorize the file so you (or the program) can identify it later. Because the three file types you are likely to use most often are the saved STAT101 worksheet (MTW), data or ASCII text files (DAT), and outfiles (LIS), this chapter explains these three in detail, and summarizes the other file types at the end.

Saving and Retrieving STAT101 Worksheets

Saved Worksheets (MTW Files)

The current worksheet consists of the data STAT101 has in its temporary memory. If you've just typed data into the computer, it resides in the current worksheet. To save an exact copy of the current worksheet on a disk, including cell contents, column names, and constants, use the heetSAVE command.

The SAVE command tells STAT101 to create an MTW file, which is a saved STAT101 worksheet. (STAT101 assigns the MTW extension to this file type, so that if you do choose to purchase MINITAB in the future, the file types will be compatible.) An MTW file is a highly efficient way to store data because STAT101 can retrieve MTW files very quickly.

Choose any file name you wish (excluding leading blanks, embedded blanks, and the symbols # and '). To save a worksheet containing data from a market analysis study, you could use the command:

⇨ STAT> SAVE 'MARKET'

STAT101 automatically saves the file as MARKET.MTW in the current directory.

If you don't type a file name after SAVE, STAT101 saves the worksheet in a file named STAT101.MTW. If you type a file extension, say MARKET.XXX, STAT101 adds that extension onto the file name instead of MTW.

If you want to save your worksheet in a directory other than your current working area, include the full *path name* within single quotation marks. A path name gives the program the information it needs to find the file, including the drive and the hierarchical sequence of directories the program will have to search through. For example, typing SAVE 'A:\SALES\MARKET' saves the file MARKET.MTW in the SALES directory of the A: drive. The backslash "\" tells your computer which level of the path you are on. See a DOS manual for information on using path names to save files in other directories.

Saving a Worksheet to a Floppy Disk

If you want to save your worksheet on a floppy disk, include the full path name within single quotation marks. For example, to save the MARKET worksheet on a floppy disk in the A: drive, use the command:

⇨ SAVE 'A:\MARKET'

Enclose both the drive and file name in single quotes. If your floppy disk has several directories on it, and you want MARKET to go in the SALES directory, use the command:

⇨ SAVE 'A:\SALES\MARKET'

STAT101 saves MARKET.MTW in the SALES directory of the disk in the A: drive. If there is no such directory, STAT101 warns you that the file name is invalid.

You do not have to type the file extension for an MTW file, either in saving or retrieving it, although you do if you use your own file extension.

Retrieving Saved Worksheets

The heetRETRIEVE command copies the data from an MTW file and places it in the current worksheet just as you left it when you saved it. (See Chapter 6, Command Reference, for more information.)

For example, if you wanted to retrieve a file named MARKET.MTW, use the command:

⇨ STAT> RETR 'MARKET'

If your file is on a disk in drive A:, then use the command:

⇨ STAT> RETR 'A:\MARKET'

This places the MARKET data in the current STAT101 worksheet, ready for you to analyze.

Retrieving a saved worksheet always erases any data from the existing worksheet before replacing it with the new worksheet, so be sure to save the existing worksheet first before opening a new one.

Retrieving Sample Data Sets

The STAT101 program disk includes a number of sample data sets that are described in Appendix B. These saved worksheet files are installed in the same directory as other STAT101 program files. If you installed STAT101 using the default install options, you only have to type **RETR** with the file name in single quotes. To retrieve the PULSE data set, for example, use the command:

⇨   STAT> RETR 'PULSE'

If you specified a different directory during installation, enter the path name for that directory when retrieving sample data sets, with the whole path enclosed in single quotes.

Saving STAT101 Worksheet Data in a Text (ASCII) File

MTW files are designed to be used only by STAT101 or the commercial version of MINITAB, and the SAVE and RETRIEVE commands work only with MTW files. In order to open data contained in STAT101 worksheets in other applications, or to use data from other applications in STAT101, you first need to save your data as an *ASCII text file,* a universal file type that stores information, using 128 standard characters that represent letters, numbers, and punctuation in the English language. Because text files use this standard ASCII character set, they can be read by almost any application.

Importing Data from Other Applications into STAT101

You can import data from other applications, like spreadsheets, databases, or word processors, into STAT101. First make sure the data have been saved as a text, or ASCII, file.

To retrieve text files created by other packages, you can use READ, SET, or INSERT (see Chapter 6, Command Reference, for information). The READ command, for example, enters ASCII characters stored in a text file one row at a time into one or more worksheet columns.

The text file should contain rows and columns of numbers, with at least one space or a comma between columns, and each row of data should be no more than 160 spaces long.

You should know how many columns the text file contains so you can tell STAT101 how many columns to use. You may also want to use the RESTART command first to clear the current worksheet and column names (which aren't deleted when you read in new data).

To import the file MYDATA.DAT into columns C1-C10, use the command:

⇨ STAT> READ 'MYDATA' C1-C10

Include the file extension if it is not DAT.

If the file contains alpha characters, if there are no spaces between columns of data, or if there are blanks for missing values, you will then need to use the READ subcommand called FORMAT to import the file. For help on using FORMAT, type **HELP READ FORMAT** and read the online help information.

Exporting Worksheet Data from STAT101 to a Text File

You can save the columns in your current worksheet to a text file, using the WRITE command. (See Chapter 6, Command Reference, for more information.) STAT101 automatically assigns text files you've created with the WRITE command the extension DAT, unless you assign one of your own. DAT files contain columns of ASCII text only; they do not include stored constants or column names.

If you omit a file name, STAT101 writes the columns to the Session screen, very close together, with no column names or row numbers. If you are writing to a file, STAT101 adjusts the format to make the data as compact as possible. The number of columns that you can put on one line varies with the data. If all columns do not fit on one line, STAT101 puts the continuation symbol & at the end of the line and continues the data onto the next line. When writing columns of unequal length to a file, STAT101 makes them equal by adding missing value symbols (*) to the shorter columns.

Once you save your STAT101 data in a text file, you can import it into almost any database, spreadsheet, or word processing package and use it there.

Combining Data from Different Files into a New File

Combining data from two or more files is easiest when the two files are already saved as text files. Suppose you have two ten-column text files, ONE.DAT and TWO.DAT, that you want to combine side-by-side into a 20-column text file, WIDEFILE.DAT. Use these commands:

⇨ STAT> READ 'ONE' C1-C10
   STAT> READ 'TWO' C11-C20
   STAT> WRITE 'WIDEFILE' C1-C20

Suppose you want to stack a ten-column text file, ONE.DAT, on top of another ten-column text file, TWO.DAT, and store the combined data as TALLFILE.DAT. Use these commands:

⇨ STAT> READ 'ONE' C1-C10
   STAT> INSERT 'TWO' C1-C10
   STAT> WRITE 'TALLFILE' C1-C10

To combine data stored in separate saved worksheet files, you need to first convert one of them to a text file, because when you open an MTW file, it always replaces the current worksheet.

Suppose you have two ten-column worksheets, ONE.MTW and TWO.MTW, that you want to combine side-by-side into a 20-column data set, WIDEFILE.MTW. Use these commands:

Command	What Command Does
STAT> RETRIEVE 'TWO'	RETRIEVE second worksheet
STAT> WRITE 'TWO' C1-C10	WRITE second worksheet to a text file
STAT> RETRIEVE 'ONE'	ONE is now in C1-C10 of worksheet
STAT> READ 'TWO' C11-C20	TWO is now in C11-C20 of worksheet
STAT> SAVE 'WIDEFILE'	WIDEFILE.MTW contains data from merged files

When STAT101 writes the second worksheet to a text file, the names of the columns and any constants in the worksheet are not retained, and will not appear in WIDEFILE.MTW.

To stack a ten-column worksheet, ONE.MTW, on top of another ten-column worksheet, TWO.MTW, and store the combined data as TALLFILE.MTW, use the following commands:

Command	What Command Does
STAT> RETRIEVE 'FILE2'	RETRIEVE second worksheet
STAT> WRITE 'FILE2' C1-C10	WRITE second worksheet to text file
STAT> RETRIEVE 'FILE1'	RETRIEVE first worksheet

```
STAT> INSERT 'FILE2' C1-C10      TWO is appended to bottom of ONE
STAT> SAVE 'TALLFILE'            TALLFILE.MTW contains stacked data
```

Saving Your STAT101 Session

In a given session you will use a number of different analysis techniques, reading the output on the screen as you go along. STAT101 does not automatically save your commands and output; you only see them on the screen. It is therefore a good practice to begin a session by starting an *outfile,* a text file that stores practically everything you see on the screen. When you open an outfile, STAT101 sends all subsequent commands and output not only to the screen, but also to a permanent text file (with the LIS extension, which stands for listing) that you can refer to when you need to recall the work you did during a given session. You can also use LIS files with any editor or word processor, and from there print a hard copy.

Begin an outfile at the beginning of the session by using the outfile command; for example:

⇨ `STAT> OUTFILE 'MYDATA'`

From this point on STAT101 sends a copy of practically everything that appears on the screen to the specified file, MYDATA.LIS. When the session is over, end the outfile either by exiting STAT101 or by using the command:

⇨ `STAT> NOOUTFILE`

If you type OUTFILE and use a file name that already contains previous session work, STAT101 appends the new output onto the end of the original LIS file.

Deleting Files

When you need to delete a file you've created in the STAT101 directory, perhaps because you need more space on your hard disk, or because you don't want an unneeded file to clutter a directory, you can use the SYSTEM command to go to the DOS prompt, and then delete it from DOS without having to stop STAT101, using the DOS DEL command. Figure 5–1 illustrates creating and then deleting an outfile called MYDATA (using the RETRIEVE and INFO commands just to put some data in the file).

FIGURE 5–1

```
STAT> OUTFILE 'MYDATA'
STAT> RETRIEVE 'CRANKSH'
 WORKSHEET SAVED 3/ 4/1993

Worksheet retrieved from file: CRANKSH.MTW
STAT> INFO

COLUMN     NAME      COUNT
C1         AtoBDist  125
C2         Month     125
C3         Day       125

CONSTANTS USED: NONE

STAT> NOOUTFILE
STAT> SYSTEM
To return to STAT101, type "EXIT" at the DOS prompt

Microsoft(R) MS-DOS(R) Version 5.00
          (C)Copyright Microsoft Corp 1981-1991.

C:\STAT101>DEL MYDATA.LIS
```

Figure 5–1 shows creating an outfile called MYDATA. STAT101 adds the extension LIS. The SYSTEM command gives access to DOS, and the DOS DEL command deletes the outfile MYDATA.LIS.

The DOS command requires path information if the directory is different from the current directory, and the full file name, including the file extension. In Fig. 5–1, the current directory is the STAT101 directory, so no additional path information was necessary.

Printing

Printing Directly to a Printer

If your computer is linked directly to a printer, you can use the PAPER command to send subsequent output to a printer. Use the command:

⇨   STAT> PAPER

Any commands or output that you enter in the Session screen will be printed on your printer until you tell it to stop by using the command:

⇨   STAT> NOPAPER

The PAPER command includes subcommands that let you control output size. See Chapter 6, Command Reference, for more information.

PAPER and OUTFILE cannot be in effect at the same time. PAPER automatically closes an outfile if one is open, and OUTFILE automatically cancels PAPER if it is in effect.

Printing to a File

If you aren't the only user connected to a printer, or if your computer has no printer at all, the best way to obtain hard copies of your session work and worksheet contents is by creating an outfile (described in the previous section) that can be read and printed by any editor or word processor. Once you have created an outfile, use STAT101's PRINT command to display worksheet data on the screen, and, simultaneously, to send it to your outfile, as shown in Fig. 5–2. (See Chapter 6, Command Reference, for more information.)

For example, to create an outfile that contains the data from the BANKING data set, start an outfile (in Fig. 5–2, this file is called BANK), and then send the worksheet contents to the screen and the file with the PRINT command.

FIGURE 5–2

```
STAT> OUTFILE 'BANK'
STAT> RETRIEVE 'BANKING'
STAT> PRINT C1 C2

 ROW    1991   1990

   1     753   1144
   2     378    351
   3     450   1002
   4     386    355
   5     314    317
   6     305    378
   7     303    335
   8     226    275
   9     171    193
  10     167    167

STAT> NOOUTFILE
```

STAT101 will create a file named BANK.LIS in text format. If you want to edit the file before you print it, you can easily do so by opening it in an editor or word processing program and printing it from there. Because the OUTFILE command saves everything in the session, the LIS file may contain much more than you actually want, or you may want to add comments here and there, so editing it before printing it is not a bad idea. You can also use OUTFILE to save only certain portions of a STAT101 session.

To print worksheet data in a more compact form, you can use the WRITE command discussed earlier in this chapter. WRITE sends the data to a text file in a compact format but does not show column names. Print this text file in the usual way. Use this method when you want to print a large volume of output efficiently.

STAT101 File Summary

Figure 5-3 summarizes STAT101's five file types and the commands that use them.

FIGURE 5-3

File Extension	Created By	Used By	Description
MTW	SAVE	RETRIEVE	Binary STAT101 worksheet. Can be retrieved only on the same computer operating system that saved it.
MTP	SAVE, PORTABLE	RETRIEVE, PORTABLE	ASCII portable STAT101 worksheet. Can be saved on one computer operating system and retrieved on a different one.
DAT	WRITE editor	READ, SET, INSERT other software	ASCII Text file. Contains data from a STAT101 worksheet. Can be edited, printed, and read by other software.
LIS	OUTFILE, NOOUTFILE	editor other software	ASCII Outfile. Contains a record of the commands you enter and the output from STAT101.
MTB	STORE data editor	EXECUTE	ASCII Macro File. Contains commands you can execute as a STAT101 macro. Can be edited, printed, and read by other software.

You can get more information on STAT101 file types by typing HELP OVERVIEW and choosing the Files and Devices option. You can also type HELP with any of the commands or subcommands associated with these file types, like HELP STORE. Chapter 6, Command Reference, also documents all the commands listed here.

6

Command Reference

The Command Reference chapter begins with a listing of STAT101 commands organized by group, with short summaries of each command's function. It then documents all commands in STAT101, in alphabetical order, including accompanying subcommands when appropriate, and examples that you can try on your own.

The examples use either data sets included with the STAT101 package or small data sets that you can enter quickly by following the instructions. If you are trying a number of examples in a row, remember that data left in the worksheet from a previous example may interfere with a current example. You may want to use the RESTART command to clear the worksheet between examples.

Command Reference examples show the actual STAT101 screen as the commands were executed, including the output. You can follow the example on your own computer by typing the same commands that are shown in the screen shot.

Command Groups

This section summarizes the tasks STAT101 can perform, giving you a place to start when you want to know which commands are appropriate for a certain task. When you've located the group of commands you need, you can turn to the alphabetic list of commands that follows for specific information and examples for each command.

Data Entry

The Data Editor is often the easiest way to enter data (see Chapter 3, Using the Data Editor), but in some cases you will enter it in the Session screen, as when you're importing data or creating random data. The following table lists different data entry types and the corresponding command or method of entry.

Entering data row by row	Data Editor or READ
Entering data a column at a time	Data Editor or SET
Generating patterned data, such as the numbers 1 through 10	SET
Inserting rows of data between established rows or adding data at the beginning or end of a column	INSERT
Indicating that all the data is entered	END
Naming columns	Data Editor or NAME
Generating random data	RANDOM

Viewing Data on the Screen

Use the Data Editor to see the columns and rows of your worksheet in a spreadsheet format.

Viewing worksheet data on the Session screen	PRINT
Viewing worksheet data on the Session screen in a compressed format	WRITE
Viewing ASCII text files on the Session screen	TYPE
Controlling the amount of output from certain statistical commands	BRIEF

T-test and confidence interval for the difference of two means; identical to TWOSAMPLE, except data for both groups is in one column with subscripts in a second column	TWOT
Correlation (Pearson) between pairs of columns	CORRELATION

Regression

Fit a regression equation to data	REGRESS. Use NOCONSTANT to fit subsequent equations without a constant (intercept) term; CONSTANT reverses effect.
Do stepwise regression	STEPWISE. Use NOCONSTANT to fit equation without a constant (intercept) term; CONSTANT reverses effect.
Control the amount of output from a regression command	BRIEF

Analysis of Variance

One-way analysis of variance with the data for each level in a separate column	AOVONEWAY
One-way analysis of variance with data in one column and subscripts in another	ONEWAY
Two-way analysis of variance for balanced data	TWOWAY

Nonparametrics

Two-sided runs test on the data to test for random order	RUNS
Sign test	STEST
Calculate sign confidence intervals	SINTERVAL
One-sample Wilcoxon signed-rank test	WTEST
Calculate confidence interval corresponding to Wilcoxon signed-rank test	WINTERVAL
Two-sample rank test and confidence interval for the difference between two population medians	MANN-WHITNEY
Generalization of Mann-Whitney test as a nonparametric alternative to the usual one-way analysis of variance	KRUSKAL-WALLIS
Walsh averages of each pair of columns	WALSH
Computes all pairwise differences	WDIFF
Pairwise slopes to calculate robust estimates of the slope of a line through the data	WSLOPE

Tables

Tally columns (one-way), with options to include counts, cumulative counts, percentages, and cumulative percentages	TALLY
Chisquare test for independence or association on a table that is already formed and stored in the worksheet	CHISQUARE
Create one-way, two-way, and multi-way tables that can include counts, percents, and basic and summary statistics	TABLE

Time Series

Compute, plot, and store the autocorrelation function (correlogram)	ACF
Compute, plot, and store the partial autocorrelation function	PACF
Compute and plot the cross correlation function	CCF
Compute differences between the elements of a column	DIFFERENCES
Move numbers in a column down a specified number of rows	LAG
Fit a non-seasonal or seasonal model to a time series	ARIMA

(continued)

Time series plot	TSPLOT
Control the amount of ARIMA output	BRIEF

Distributions and Random Data

Generate columns of random data, optionally using a specified distribution	RANDOM. BASE sets a starting point so you can repeat the generation.
Compute discrete probabilities, or the probability density function, for the given values from a specified distribution	PDF
Compute cumulative probabilities for given values from a specified distribution	CDF
Compute the inverse of the cumulative distribution function. "Look up" the critical values for a distribution	INVCDF
Randomly sample rows from a column, with or without replacement	SAMPLE

Miscellaneous Commands

Start STAT101	Type **STAT101** after the DOS prompt.
Clear the current worksheet	RESTART
Stop STAT101	STOP
Repeat the last command line	[F3]

Switch back and forth between Session screen and Data Editor	[Esc] key or [Alt] + [D] to move to the Data Editor and [Alt] + [M] to move back
Get help	HELP in the main screen. In the Data Editor, press [F1].
Worksheet summary	INFO
Temporarily access DOS	SYSTEM
Cancel a command	ABORT as a subcommand
Cancel the display of output on the screen	[Ctrl] + [C]
Control the display of the Continue? prompt	OH 0 suppresses the prompt. The default is OH 24.

Macros

Store all commands that follow in a file that can be executed using the EXECUTE command; useful for saving a set of commands used frequently.	STORE to start storing commands and END to stop
Execute a macro, or file of stored commands.	EXECUTE
When you execute a file, commands are shown (echoed) on the Session screen.	NOECHO keeps commands from a stored file from being printed on the screen; ECHO restores printing of commands
Cancel a macro	[Ctrl] + [C]

Command Syntax

The command reference list that follows gives each command with its syntax (see Chapter 4, Issuing Commands, for information on using STAT101 command syntax). The Command Reference uses the following conventions for worksheet and command elements, including arguments:

C	denotes a column, such as C12 or 'Height'
K	denotes a constant, such as 8.3 or K14
E	denotes either a constant or a column
[ ]	encloses an optional argument
CAPITALS	denote a STAT101 command or subcommand, such as SAVE

* Missing Value Symbol

How to Use

STAT101 uses an asterisk, *, to represent missing values in a column of numbers. In alpha columns (columns with letters or "strings" instead of numbers) a blank is used.

You can substitute an * for a missing or invalid value when entering data with READ, SET, or INSERT, as shown in Fig. 6–1.

FIGURE 6–1
```
STAT> READ C1-C3
DATA>    0 3 *
DATA>    9 * 5
DATA> END
```

Many STAT101 commands can operate on * if it is enclosed in single quotes, as shown in Fig. 6–2, which shows replacing -99 values with missing values, and then copying the original -99 values into C3.

FIGURE 6–2

```
STAT> SET C1
DATA>    5 2 -99 3 -99
DATA> END
STAT> CODE (-99) '*' C1 C2
STAT> COPY C1 C3;
SUBC>    USE C2 = '*'.
STAT> PRINT C1-C3

  ROW     C1    C2    C3
    1      5     5   -99
    2      2     2   -99
    3    -99     *
    4      3     3
    5    -99     *
```

STAT101 operations skip over missing values as necessary. For example, the PLOT command plots only points that have both an x and a y coordinate. REGRESS only uses cases with the dependent and all the independent variables present. When an arithmetic command (e.g., ADD, LOG, SQRT) operates on a missing value, the answer will be a missing value.

Comment Symbol

How to Use

STAT101 ignores everything you type between a # symbol and the end of a line. You can use this symbol to include comments on command lines, subcommand lines, data lines, or lines by themselves, as shown in Fig. 6–3.

FIGURE 6–3

```
STAT> OUTFILE 'A:\HOMEWORK'
STAT> #Name: Joan Carey
STAT> #Professor:  Dr. Gillett
STAT> RETR 'SALES'
STAT> PRINT C1-C4      #print sales revenue for 1988-1991
STAT> NOOUTFILE
```

& Continuation Symbol

How to Use

You can type an & and then [Enter] at the end of any line to indicate that the command or data row continues on to the next line. STAT101 will return with a CONT> prompt, as shown in Fig. 6–4 on the following page.

Figure 6-4

```
STAT> REGRESS the data in column C1 on        &
CONT> 1 independent variable in C4            &
CONT> store residuals in C10, fits in C11
```

Press [Enter] after the last continuation line to execute the command.

ABSOLUTE

Calculates absolute values, changing all negative numbers to positive numbers.

Syntax

ABSOLUTE   value of **E**, put into **E**

Also see Arithmetic and Transformations.

ACF

Displays autocorrelations of time-series data in a graph.

Syntax

ACF   [up to **K** lags] for data in **C** [put into **C**]

How to Use

This command computes the autocorrelation of the time series for time lags varying from 1 to K. (If you don't specify K, STAT101 uses the value $\sqrt{n} + 10$, where n is the number of observations in the series.) STAT101 plots the autocorrelation values in a correlogram and optionally stores them in a new column.

Figure 6-5 is an ACF of the number of gallons of water used each month from 1983 through June, 1986.

FIGURE 6-5

```
STAT> RETRIEVE 'UTILITY'
STAT> ACF 'GAL'

ACF of GAL
            -1.0 -0.8 -0.6 -0.4 -0.2  0.0  0.2  0.4  0.6  0.8  1.0
            +----+----+----+----+----+----+----+----+----+----+
    1 -0.063                         XXX
    2  0.041                         XX
    3 -0.087                         XXX
    4 -0.098                         XXX
    5 -0.125                         XXXX
    6 -0.038                         XX
    7 -0.197                       XXXXXX
    8  0.015                         X
    9 -0.093                         XXX
   10  0.018                         X
   11 -0.047                         XX
   12  0.190                         XXXXXX
   13  0.003                         X
   14 -0.028                         XX
   15  0.151                         XXXXX
   16 -0.011                         X
```

Also see PACF, CCF, and LAG.

ACOS

Calculates arccosines. Give arguments in radians; STAT101 returns answers in radians. To convert radians to degrees, multiply by 57.297. If an argument is not between -1 and +1, STAT101 sets the answer equal to * (missing value symbol).

Syntax

ACOS	(arccosine) of **E**, put into **E**

Also see Arithmetic and Transformations.

ADD

Does row-by-row addition on columns and/or constants. If you add columns of unequal length, STAT101 uses the length of the shortest column as the length of the answer column.

Syntax

ADD   E to E ... to E, put into E

Also see Arithmetic and Transformations.

ANTILOG

Raises 10 to the power E.

Syntax

ANTILOG   (10 to the power) of E, put into E

Also see Arithmetic and Transformations.

AOVONEWAY

Computes a one-way analysis of variance between two or more columns.

Syntax

AOVONEWAY   the data in C ... C

How to Use

This command computes a one-way analysis of variance for two or more columns of data. AOVONEWAY assumes that each column contains sample data drawn from possibly different populations with a common variance, σ^2. The sample size in each column may be different. (If you specify only two columns, AOVONEWAY is equivalent to the TWOSAMPLE command with the POOLED subcommand.)

Figure 6–6 compares how car speeds were affected by: Method 1) posting a speed limit sign; Method 2) Method 1 plus a "Speed Limit Strictly Enforced" sign; Method 3) Method 1 plus a "Speed Limit Radar Controlled" sign; and Method 4) Method 1 plus a parked, empty patrol car.

FIGURE 6-6

```
STAT> RETRIEVE 'CARSPEED'
STAT> AOVONEWAY 'METHOD 1' 'METHOD 2' 'METHOD 3' 'METHOD 4'

ANALYSIS OF VARIANCE
SOURCE      DF       SS        MS       F       p
FACTOR       3    555.4     185.1    5.12   0.004
ERROR       40   1447.7      36.2
TOTAL       43   2003.2
                                     INDIVIDUAL 95 PCT CI'S FOR MEAN
                                     BASED ON POOLED STDEV
LEVEL       N     MEAN     STDEV   --------+---------+---------+---------
METHOD 1   12   64.417    7.103                         (------*------)
METHOD 2   10   61.700    6.961                    (------*-------)
METHOD 3   11   57.727    5.081              (------*-------)
METHOD 4   11   55.364    4.456      (--------*------)
                                    --------+---------+---------+---------
POOLED STDEV =   6.016               55.0      60.0      65.0
```

Also see ONEWAY, TWOSAMPLE, and KRUSKAL-WALLIS.

ARIMA

Fits seasonal or nonseasonal models to a time series, and lets you predict future values.

Syntax

ARIMA	p = **K** d = **K** q = **K** [P = **K** D = **K** Q = **K** S = **K**] for data in **C** [put residuals into **C** [predicted values into **C** [coefficients into **C**]]]
CONSTANT	
NOCONSTANT	
STARTING	
FORECAST	[values in **C**] up to **K** leads [store forecasts in **C** [confidence limits in **C C**]]

How to Use

This command permits you to fit time-series data with an autoregressive integrated moving average (ARIMA) model, and optionally, to correct for seasonal trends. The input consists of a column of time-series data, plus parameters on the command line that indicate the exact model to be used. The FORECAST subcommand predicts future values for the same series. You can store results in the worksheet for future use, and you can increase or decrease the amount of information printed, using the BRIEF command.

For nonseasonal data, the model is fully specified by the parameters p, d, and q, where p is the order of the autoregressive (AR) component of the model, d is the number of differences used to discount trends over time, and q is the order of the moving-average (MA) component.

For example, the following command fits a moving average model of order 1 to a column of data:

⇨ STAT> ARIMA 0 0 1 C2

To describe recurring seasonal trends within the time series, the additional parameters P, D, Q, and S are required, where S is the number of units in a complete cycle. For example, for monthly data showing a one-year pattern of recurrence, set S = 12. The other terms are analogous to p, d, and q for the nonseasonal part: P is the order of the seasonal autoregressive component, D is the number of seasonal differences, and Q is the order of the seasonal moving-average component.

ARIMA allows you to save the residual values in the STAT101 worksheet by naming an additional column on the command line. If you name two columns, the residuals as well as the predicted values are stored. Finally, a third storage column can hold the estimated model coefficients.

Figure 6–7 fits a seasonal time-series model to analyze quarterly electricity costs.

FIGURE 6-7

```
STAT> RETRIEVE 'UTILITY'
STAT> ARIMA (0 0 1) (0 1 1) 4 'KWH $'
```

Estimates at each iteration
Iteration	SSE	Parameters	
0	177091	0.100	0.100
1	129217	-0.034	0.250
2	99214	-0.137	0.400
3	79415	-0.211	0.550
4	65288	-0.268	0.700
5	55414	-0.321	0.850
6	51011	-0.471	0.863
7	48190	-0.621	0.870
8	46619	-0.735	0.870
9	45885	-0.807	0.866
10	45676	-0.843	0.857
11	45663	-0.853	0.853
12	45662	-0.853	0.854

Relative change in each estimate less than 0.0010

Final Estimates of Parameters
Type		Estimate	St. Dev.	t-ratio
MA	1	-0.8532	0.1033	-8.26
SMA	4	0.8538	0.1743	4.90

Differencing: 0 regular, 1 seasonal of order 4
No. of obs.: Original series 44, after differencing 40
Residuals: SS = 44198.5 (backforecasts excluded)
MS = 1163.1 DF = 38

Modified Box-Pierce chisquare statistic
Lag	12	24	36	48
Chisquare	25.7(DF=10)	45.7(DF=22)	49.3(DF=34)	* (DF= *)

ARIMA Specifications

1. In the parameter list, at least one of the p/P or q/Q parameters must be nonzero, and none may exceed 5.
2. The maximum number of parameters to be estimated is 10.
3. There must be at least 3 data points left after differencing. That is, $S * D + d + 2$ must be less than the number of points.
4. The maximum "back order" for the model is 100. In practice, this condition is always satisfied if $S * D + d + p + P + q + Q$ is at most 100.
5. The ARIMA model normally includes a constant term only if there is no differencing (i.e., $d = D = 0$). See the CONSTANT and NOCONSTANT subcommands.

6. Missing observations are only allowed at the beginning or the end of a series, not in the middle.
7. The seasonal component of this model is multiplicative, and thus is appropriate when the amount of cyclical variation is proportional to the mean.

Subcommands

CONSTANT

NOCONSTANT

These subcommands explicitly instruct STAT101 whether or not to include a constant term in the model. NOCONSTANT is the default condition if either d or D is positive.

STARTING   values in C

You can specify starting values for the parameters in the ARIMA model. They must appear in a column in the same order as on the output: p (AR values), P (seasonal AR values), q (MA values), Q (seasonal MA values), and an optional constant. Otherwise STAT101 uses the initial value 0.1 for each parameter except the constant.

FORECAST   [forecast origin = K] up to K leads [store forecasts in C [confidence intervals in C C]]

This subcommand allows you to predict the next K observations for the series. If the origin is not specified, it is assumed to be the end of the current series (i.e., the forecasts are for the future). Forecast values and confidence limits can be stored and plotted at a future time.

Also see ACF, PACF, LAG, and BRIEF.

Arithmetic and Transformations

Perform arithmetic and mathematical transformations on columns, stored constants, and numbers.

How to Use

Because the commands that perform mathematical operations work in a similar manner, they are grouped together here under the following headings: LET, arithmetic commands, column functions, and row functions.

LET

The LET command performs many different operations on values you specify. As part of the expression, you can enter arithmetic, comparison, or logical operators, or a variety of functions including logarithms, standard deviations, and square roots.

You can assign new values to one element in a column, to a constant, or to an entire column with the LET command. Its form is the following, with no extra text allowed (except after #):

LET   **E** = expression

To change one element in a column—for example, to replace the third row in C1 with a 5—you can use the command:

⇨ `STAT> LET C1(3) = 5`

You can also use a stored constant to indicate the row number:

⇨ `STAT> LET K1 = 3`
    `STAT> LET C1(K1) = 5`

In addition, you may use SORT, RANK, LAG, and any of the arithmetic or column functions in a LET expression if you include parentheses around its argument. For example, type SQRT(C1), not SQRT C1.

Within parentheses, evaluation follows the usual order of precedence: functions are always evaluated first, followed by exponentiation, then multiplication and division, and finally addition and subtraction. When two or more operations have the same precedence, they are done from left to right.

This simple example

⇨ STAT> LET C5 = C1 + C2

stores the sum of the numbers in columns C1 and C2 in C5.

The more complex command

⇨ STAT> LET K2 = STDEV(C5)**2

computes the square of the standard deviation (i.e., the variance) of C5, and stores it in the constant K2.

STAT101 uses the following symbols:

**	raise to a power
*	multiplication
/	division
+	addition
-	subtraction
= or EQ	equal to
~= or NE	not equal to
< or LT	less than
> or GT	greater than
<= or LE	less than or equal to
>= or GE	greater than or equal to
& or AND	logical "and"
| or OR	logical "or"
~ or NOT	logical "not"

If any element of a row is missing, LET sets the result to *, the missing value symbol. If an operation is impossible, like dividing by 0, LET sets that result to missing and prints a "value out of bounds" message. LET sets true comparisons to 1 and false comparisons to 0. You can use the logical comparison operators to subset your data.

Arithmetic Functions

You can use the following arithmetic operations as functions within a LET command, as described above, or as standalone commands. As independent commands they take the forms shown below:

ADD	E to E . . . to E,	put into E
SUBTRACT	E from E,	put into E
MULTIPLY	E by E . . . by E,	put into E
DIVIDE	E by E,	put into E
RAISE	E to the power of E,	put into E
ABSOLUTE	value of E,	put into E
SIGNS	of E,	put into E
SQRT	of E,	put into E
ROUND	E to integer,	put into E
LOGE	(natural logarithms) of E,	put into E
LOGTEN	(logarithms, base 10) of E,	put into E
EXPONENTIATE	(e to the power) of E,	put into E
ANTILOG	(10 to the power) of E,	put into E
NSCORE	(normal scores) of C,	put into C
SIN	of E,	put into E

(continued)

COS	of E,	put into E
TAN	of E,	put into E
ASIN	(arcsine) of E,	put into E
ACOS	(arccosine) of E,	put into E
ATAN	(arctangent) of E,	put into E
RANK	the values in C,	put into C
LAG	[of length K] the values in C,	put into C
PARSUMS	(partial sums) of C,	put into C
PARPRODUCTS	(partial products) of C,	put into C

Figure 6–8 shows examples of these functions.

FIGURE 6–8
```
STAT> READ C1-C3
DATA>   0  4  8
DATA>  -1 10  0
DATA>   *  6 -1
DATA>   3  7  *
DATA> END
STAT> ADD C1 C2 3 C5
STAT> PARSUMS C1 C6
STAT> RAISE C1 2 C7
STAT> LET C8 = SQRT(C2) + SIN(C1)
STAT> SIGNS C1 C9
        1  NEGATIVE VALUES     1  ZERO VALUES      1  POSITIVE VALUES
STAT> PRINT C1 C2 C5-C9

  ROW   C1    C2    C5    C6   C7        C8     C9

   1     0     4     7     0    0    2.00000    0
   2    -1    10    12    -1    1    2.32081   -1
   3     *     6     *     *    *         *     *
   4     3     7    13     *    9    2.78687    1
```

The SIGNS command converts negative numbers, zero, and positive numbers to −1, 0, and +1 respectively, and optionally stores these new values in a column.

Column Functions

The column functions each compute one number from a column of data and can be used in LET expressions. As standalone commands they have the following form:

COUNT	total number of rows	in **C** [put in **K**]
N	number of nonmissing values	in **C** [put in **K**]
NMISS	number of missing values	in **C** [put in **K**]
SUM	add all values	in **C** [put in **K**]
MEAN	average all values	in **C** [put in **K**]
STDEV	standard deviation	of **C** [put in **K**]
MEDIAN	middle data value	of **C** [put in **K**]
MINIMUM	smallest number	in **C** [put in **K**]
MAXIMUM	largest number	in **C** [put in **K**]
SSQ	sum of the squared values	in **C** [put in **K**]

Figure 6–9 uses the same data as Fig. 6–8.

FIGURE 6–9

```
STAT> MEAN C1 K1

   MEAN   =      0.66667

STAT> LET C5 = COUNT(C1) - N(C1)
STAT> PRINT K1 C1 C5

K1         0.666667

   ROW    C1    C5

    1      0     1
    2     -1
    3      *
    4      3
```

Arithmetic and Transformations

Row Functions

The row functions shown below compute one number for each row in a set of columns. You can only use them as standalone commands, and the answers are always stored in a column. You can also use a stored constant or number with the row functions.

RCOUNT	total number of values per row of E . . . E, put in C
RN	number of nonmissing values per row of E . . . E, put in C
RNMISS	number of missing values per row of E . . . E, put in C
RSUM	sum of all values per row of E . . . E, put in C
RMEAN	average of all values per row of E . . . E, put in C
RSTDEV	standard deviation of all values per row of E . . . E, put in C
RMEDIAN	middle data value per row of E . . . E, put in C
RMINIMUM	smallest number per row of E . . . E, put in C
RMAXIMUM	largest number per row of E . . . E, put in C
RSSQ	sum of the squared values per row of E . . . E, put in C

Figure 6–10 illustrates this, using the same data used in Fig. 6–8 and Fig. 6–9.

FIGURE 6–10
```
STAT> RMEAN C1-C3 C5
STAT> RMINIMUM C1-C3 C6
STAT> RMAXIMUM C1-C3 C7
STAT> RSUM 10 20 C1 C8
STAT> PRINT C1-C3 C5-C8

ROW   C1    C2    C3    C5    C6    C7    C8

 1     0     4     8    4.0    0     8    30
 2    -1    10     0    3.0   -1    10    29
 3     *     6    -1    2.5   -1     6    30
 4     3     7     *    5.0    3     7    33
```

ASIN

Calculates arcsines. Give arguments in radians; STAT101 returns answers in radians. To convert radians to degrees, multiply by 57.297. If an argument is not between -1 and +1, STAT101 sets the answer equal to * (missing value symbol).

Syntax

ASIN (arcsine)   of E, put into E

Also see Arithmetic and Transformations.

ATAN

Calculates arctangents. Give arguments in radians; STAT101 returns answers in radians. To convert radians to degrees, multiply by 57.297. If an argument is not between -1 and +1, STAT101 sets the answer equal to * (missing value symbol).

Syntax

ATAN (arctangent)   of E, put into E

Also see Arithmetic and Transformations.

BASE

Fixes a starting point for STAT101's random number generator.

Syntax

BASE	K

How to Use

The command RANDOM allows you to generate any amount of new data having the statistical distribution you specify. STAT101 normally chooses its own starting point for this process so that RANDOM generates different data each time.

Sometimes, however, you may want to generate the same data more than once without having to save it in a file. In that case, your first RANDOM command should be preceded by the command BASE, together with a number of your choice, telling the random number generator how to begin. You can then produce identical sample data any time you repeat the same BASE and RANDOM commands.

Also see RANDOM.

BOXPLOT

Depicts a column of data with a "box and whisker" display.

Syntax

BOXPLOT	for data in C
START	at K [end at K]
INCREMENT	= K
NOTCH	[K% confidence] sign confidence interval
BY	C

LEVELS	K . . . K [for C]
LINES	= K

How to Use

Boxplots, like histograms and stem-and-leaf displays, give an overview of the values in a single column and allow you to identify any extreme values. This command can also depict a confidence interval for the population median.

The display consists primarily of a rectangular box, representing the middle half of the data, and dashed lines, or "whiskers," extending to either side, indicating the extent of the data. The median value is marked with a tick mark (+) inside the box, and the left and right ends of the box, each marked with an "I", are called hinges and essentially represent the first and third quartiles of the data.

The whiskers extend to the left and right to the last observation on either side within one and one-half times the box width from the box ends. Any value lying between one and one-half and three times the box width beyond the box ends is considered a possible outlier and is plotted with an *. Any more remote value is considered an extreme outlier and is plotted with a 0.

The selling prices of homes with and without a basement are displayed in Fig. 6–11.

FIGURE 6–11

```
STAT> RETRIEVE 'REALEST'
STAT> BOXPLOT 'SELL $';
SUBC>    NOTCH;
SUBC>    BY 'BASEMENT'.
```

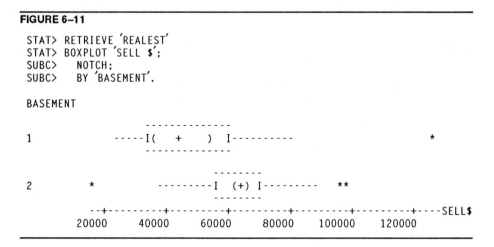

Subcommands

START	at K [end at K]

This subcommand specifies the first and optionally the last tick mark on the axis. STAT101 excludes points falling outside this range.

INCREMENT = K

This subcommand determines the distance between the tick marks (+ symbols) on the horizontal axis.

NOTCH   [K percent confidence] sign confidence interval

When this subcommand is included, parentheses (notches) in each boxplot indicate the limits of a sign confidence interval for the population median, calculated as for the SINT command. STAT101 finds a 90% confidence interval if no other value is given.

STAT101 uses a priority system to determine what symbols will be displayed in the BOXPLOT output. If the median falls on the same space as a notch (CI parenthesis), the notch is not displayed. Similarly, if the median and an I-hinge fall on the same space, the I is not displayed.

BY   C

This subcommand allows you to create separate boxplots for rows of data categorized by their values in another column C, which should contain only integers between −1000 and +1000 or the missing value symbol (*).

LEVELS   K ... K

STAT101 uses the LEVELS option only in coordination with the subcommand BY when you wish to rearrange or omit the displays for some subgroups, drawing one boxplot for each of the BY subscript values (levels) in the order you list them.

LINES = K

The option LINES = 1 horizontally condenses each boxplot from three lines to one.

Also see STEM-AND-LEAF and SINTERVAL.

BRIEF

Controls the amount of output for the REGRESS and ARIMA commands.

Syntax

BRIEF   K

How to Use

This command is used before either REGRESS or ARIMA to control the amount of information printed. The larger the value of K, the more output. BRIEF 2 is the level of detail normally provided.

The following output is available for the command REGRESS:

K = 0   No output is printed. All requested storage is done.

K = 1   The regression equation, table of coefficients, s, R-squared, R-squared adjusted, and the first portion of the analysis of variance table (regression, error, and total sum of squares).

K = 2   In addition, the second portion of the analysis of variance table, listing each predictor variable and its sequential sum of squares, and unusual observations (those marked with an X or an R—see the REGRESS command).

K = 3   In addition, the full table of fitted and residual values.

The following output is available for the command ARIMA:

K = 1   The table of final estimates, differencing information, residual sum of squares, and the number of observations.

K = 2   In addition, a table of estimates at each iteration, and the backforecasts if they are not dying out rapidly.

K = 3   In addition, the correlation matrix of the estimated parameters.

K = 4   In addition, the backforecasts are always printed.

Also see ARIMA and REGRESS.

CCF

Displays the cross correlations for time-series data in a graph.

Syntax

CCF   [up to **K** lags] between data in **C** and **C**

How to Use

This command computes and plots the cross correlations between two time series for time lags varying from −K to K. The default value for K is $\sqrt{n}$ + 10, where n is the number of rows.

In Fig. 6–12 the commands cross correlate river temperatures, taken hourly upstream from a power plant, with the pH value of the water. (Only a portion of the output is displayed.)

FIGURE 6–12

```
STAT> RETRIEVE 'RIVER2'
STAT> CCF 24 'TEMP' 'PH'

CCF - correlates TEMP(t) and PH(t+k)

              -1.0 -0.8 -0.6 -0.4 -0.2  0.0  0.2  0.4  0.6  0.8  1.0
              +----+----+----+----+----+----+----+----+----+----+
   -24  0.381                              XXXXXXXXXX
   -23  0.331                              XXXXXXXXX
   -22  0.231                              XXXXXXX
   -21  0.093                              XXX
   -20 -0.053                             XX
   -19 -0.201                         XXXXXX
   -18 -0.303                      XXXXXXXXX
   -17 -0.392                   XXXXXXXXXXX
   -16 -0.489                 XXXXXXXXXXXXX
   -15 -0.555               XXXXXXXXXXXXXXX
   -14 -0.590               XXXXXXXXXXXXXXX
   -13 -0.585               XXXXXXXXXXXXXXX
   -12 -0.539                XXXXXXXXXXXXX
   -11 -0.473                  XXXXXXXXXXX
   -10 -0.377                     XXXXXXXXX
    -9 -0.229                        XXXXXX
    -8 -0.076                            XXX
    -7  0.071                              XXX
    -6  0.242                              XXXXXXX
```

Also see ACF, PACF, and LAG.

CD

Changes to a different directory or shows the current directory.

Syntax

CD	[path]

How to Use

CD works in STAT101 just as it does in the DOS operating system. For example, to change your current working directory to C:\STAT101\WORK, use the command:

⇨ STAT> CD \STAT101\WORK

CD without a path displays the current directory. See the DOS manual that came with your computer for more information.

Also see DIR, TYPE, and SYSTEM.

CDF

Computes cumulative probabilities for the given values from a specified distribution.

Syntax

CDF	for values in **E** [put into **E**]
BINOMIAL	n = **K**, p = **K**
CHISQUARE	degrees of freedom = **K**
DISCRETE	values in **C**, probabilities in **C**
F	df numerator = **K**, df denominator = **K**
INTEGER	discrete uniform distribution on **K** to **K**

(continued)

NORMAL	[mean = **K** [standard deviation = **K**]]
POISSON	mean = **K**
T	degrees of freedom = **K**
UNIFORM	continuous distribution [on **K** to **K**]

How to Use

The cumulative distribution function (CDF) for a random variable gives the probability that its value is less than or equal to x. The CDF command computes this probability for a constant or for each number in a column, as shown in Fig. 6–13.

FIGURE 6–13

```
STAT> CDF 75;
SUBC>    NORMAL 77 2.

   75.0000     0.1587
```

If you omit a subcommand, STAT101 uses a normal distribution with a mean of zero and standard deviation of one.

For discrete distributions only (BINOMIAL, POISSON, INTEGER and DISCRETE), you may omit the column or constant on the main command line to have STAT101 produce a full table of cumulative probabilities.

Subcommands

See the listing under the command RANDOM for details on CDF subcommands.

Also see RANDOM, PDF, and INVCDF.

CHISQUARE

Does a chisquare test of association for the table of counts given in the input columns.

Syntax

CHISQUARE analysis of the table in C ... C

How to Use

This command computes a chisquare test of association (non-independence) on a contingency table composed of two to seven columns from the worksheet. STAT101 prints a table of observed values and expected values, the total chisquare statistic as the sum of its components from each cell of the table, and the number of degrees of freedom. STAT101 also lists the number of cells with small expected frequencies (less than 5), if any.

Suppose you want to test whether brake failure occurs more often in cars purchased from dealers or privately. In Fig. 6–14, there were 931 defective brakes and 2723 good brakes in cars bought from dealers, with 1690 defective and 3498 good brakes in cars bought privately. Row 1 refers to the number of defective brakes found, while row 2 is the number of good brakes found.

FIGURE 6–14

```
STAT> READ C1 C2
DATA>    931   1690
DATA>   2723   3498
DATA> END
STAT> NAME C1 'DEALER' C2 'PRIVATE'
STAT> CHISQUARE C1 C2

Expected counts are printed below observed counts

          DEALER   PRIVATE    Total
     1       931      1690     2621
          1083.14   1537.86

     2      2723      3498     6221
          2570.86   3650.14

Total      3654      5188     8842

ChiSq = 21.370 +  15.051 +
         9.004 +   6.341 = 51.767
df = 1
```

Also see **TABLE** with the CHISQUARE subcommand.

CODE

Copies one or more columns while recoding values.

Syntax

CODE   (K...K) to K...(K...K) to K for data in C...C, put into C...C

How to Use

With this command, you can recode any values you choose as you copy data into a new set of columns, as shown in Fig. 6–15.

FIGURE 6–15

```
STAT> SET C1
DATA>   -99  3  8  4  -99  0
DATA> END
STAT> CODE (-99) '*' (1:5)1 (6:10)2 C1 C2
STAT> PRINT C1 C2

 ROW    C1    C2
  1    -99    *
  2     3     1
  3     8     2
  4     4     1
  5    -99    *
  6     0     0
```

In Fig. 6–15, the CODE command copies column C1 to C2, changing all occurrences of –99 to *, numbers 1 through 5 to 1, and numbers 6 through 10 to 2.

Enclose each list or range of values to be changed in parentheses, followed by the new code. Then list the input columns, and an equal number of output columns (up to 15 of each). If you use an * on the command line to code missing values, enclose it in single quotes.

You may also use stored constants in place of numbers:

⇨   STAT> CODE (1:K1) K2 C1 C2

Also see COPY.

CONCATENATE

Joins two or more alpha columns into a new, wider column.

Syntax

CONCATENATE   C . . . C, put into C

How to Use

This command pastes a set of columns together side by side, and stores them in one new column. The old columns must contain alpha data (i.e., strings of letters or other symbols); the new column will also be an alpha column, up to 80 characters wide. Unlike most commands, CONCATENATE requires that the storage column (the last column on the command) be different than the input columns.

Figure 6–16 illustrates concatenating two columns of the TEST data set that contain last and first names into a new column, C6.

FIGURE 6–16

```
STAT> RETRIEVE 'TEST'
STAT> INFO

     COLUMN    NAME      COUNT
  A  C1        LAST NAM    24
  A  C2        FIRST       24
     C3        TEST1       24
     C4        TEST2       24
     C5        TEST3       24

CONSTANTS USED: NONE

STAT> CONCATENATE C1 C2 C6
```

Use the PRINT command to see C6, the column containing the linked alpha data.

Also see READ, SET, PRINT, and CONVERT.

CONSTANT

Fits a constant term in all the REGRESS and STEPWISE models that follow.

Syntax

CONSTANT

How to Use

By default, STAT101 includes a constant term (the y-intercept of the regression line) in all REGRESS and STEPWISE models. The NOCONSTANT command (or subcommand) causes STAT101 to exclude the constant term; CONSTANT tells STAT101 to resume its usual practice.

Also see NOCONSTANT, REGRESS, and STEPWISE.

CONVERT

Converts numeric data to alpha data, or alpha data to numeric data.

Syntax

CONVERT	using the conversion table in **C** and **C**, convert the values in **C** put into **C**

How to Use

This command codes a column of alpha data to a column of numbers, or the opposite. To begin with, you need two worksheet columns to serve as a conversion table, one listing alpha expressions and one listing the corresponding numbers. These are listed on the CONVERT command line in the same order as the conversion. That is, if you're converting alpha data to numeric, list the alpha column first. If you're converting numeric data to alpha, list the numeric column first.

For example, retrieve the EXAMPLE.MTW saved worksheet and code the alpha expressions NO and YES in the column DIET to 0's and 1's and place the numbers in a new column C10. The EXAMPLE.MTW worksheet already has a conversion table in columns C6 and C7, which is shown in Fig. 6–17. (You can also press [Esc] to look at the worksheet in the Data Editor.)

FIGURE 6-17

```
STAT> RETRIEVE 'EXAMPLE'
STAT> PRINT C6 C7

ROW     C6      C7

 1      NO      0
 2      YES     1
```

Now do the conversion, as shown in Fig. 6–18.

FIGURE 6-18

```
STAT> CONVERT C6 C7 'DIET' C12
STAT> PRINT 'DIET' C12

ROW    DIET    C12

 1     NO       0
 2     YES      1
 3     YES      1
 4     YES      1
 5     NO       0
 6     NO       0
 7     YES      1
```

Also see CONCATENATE, READ, SET, PRINT, and WRITE.

COPY

Copies data to and from columns and constants.

Syntax

COPY	from C . . . C into C . . . C
USE	rows K . . . K
USE	rows where C = K . . . K
OMIT	rows K . . . K
OMIT	rows where C = K . . . K

The COPY command can also take the form:

COPY	from constants K ... K into K ... K
COPY	from C into constants K ... K
COPY	from constants K ... K into C

How to Use

This command copies all or part of the data from a set of columns or stored constants. USE says which rows to copy; OMIT says which not to copy. Both subcommands are shown in Fig. 6–19.

FIGURE 6–19

```
STAT> RETRIEVE 'EXAMPLE'
STAT> COPY 'HT' 'WT' C20 C21;
SUBC>    USE 'SEX'=1.
STAT> COPY 'HT' 'WT' C22 C23;
SUBC>    OMIT 'WT' = '*'.
STAT> COPY 'HT' C24;
SUBC>    USE 'HT' = 65:70.
STAT> PRINT 'SEX' 'HT' 'WT' C20-C24
```

ROW	SEX	HT	WT	C20	C21	C22	C23	C24
1	1	70	165	70	165	70	165	70
2	2	62	110	64	*	62	110	66
3	1	64	*	71	183	71	183	68
4	1	71	183	68	158	68	158	
5	2	66	*			61	106	
6	1	68	158					
7	2	61	106					

The first COPY command in Fig. 6–19 copies all the male ('SEX' = 1) heights and weights into C20 and C21. The second command copies all the non-missing heights and weights into C22 and C23. The last command copies all the heights from 65 to 70 into C24.

This command can also copy a set of constants into a single column, or rows of a column into the same number of stored constants. For example:

⇨ STAT> COPY K1-K3 C1

copies K1, K2, and K3 into column C1.

Subcommands

USE	rows where C = K ... K

This subcommand tells STAT101 which rows to copy. For example, to copy rows 5, 6, 7, and 8 of C1 into C2, you could use the command:

⇨
```
STAT> COPY C1 C2;
SUBC>   USE 5:8.
```

OMIT	rows where C = K ... K

This subcommand works like USE, except that it tells STAT101 which rows not to copy.

Also see LET, DELETE, and ERASE.

CORRELATE

Correlates pairs of columns.

Syntax

CORRELATE	the data in C ... C

How to Use

If you give two columns as input, CORRELATE calculates the (Pearson product moment) correlation coefficient for the pair. If you list more than two columns, the correlation is calculated for every possible pair of columns. The lower triangle of the symmetric correlation matrix is printed. See Fig. 6–20 on the following page.

FIGURE 6–20

```
STAT> RETRIEVE 'STA261'
STAT> CORRELATE 'T1' 'T2' 'T3' 'FINAL'

           T1      T2      T3
T2       0.613
T3       0.612   0.699
FINAL    0.672   0.746   0.821
```

The correlation calculation skips any pair of values that have one or both values missing. This is often called "pairwise deletion" of missing values. This method is best for individual correlation, but the correlation matrix as a whole may not be well behaved (for example, it may not be positive definite).

To calculate Spearman's rho (rank correlation) between pairs of columns of nonmissing data, rank both columns, using the RANK command, and then correlate the ranks. (Delete rows containing missing data before ranking columns.) To do this, you could use the commands:

⇨ STAT> RANK 'T1' C21
  STAT> RANK 'T2' C22
  STAT> CORRELATE C21 C22

Also see REGRESS and RANK.

COSINE

Calculates cosines. The argument must be in radians. To convert from degrees to radians, multiply by 0.017453. If the numbers are too large to be accurate, STAT101 sets the result to *, the missing value symbol, and prints a "value out of bounds" message. The exact magnitude of the out of bounds range depends on the computer.

Syntax

COSINE	of E, put into E

Also see Arithmetic and Transformations.

COUNT

Counts the number of values in a column, including missing values.

Syntax

COUNT	the number of values in **C** [put into **K**]

Also see Arithmetic and Transformations.

DELETE

Deletes rows from columns in the worksheet.

Syntax

DELETE	rows **K** ... **K** in columns **C** ... **C**

How to Use

DELETE removes selected rows from columns in the current worksheet and adjusts the remaining rows to fill in any gaps. Figure 6–21 shows the commands that change the worksheet from the Before to the After as shown in Fig. 6–22.

FIGURE 6–21

```
STAT> RETRIEVE 'EXAMPLE'
STAT> PRINT C3-C5
STAT> DELETE 2:4 C3-C5
STAT> PRINT C3-C5
```

STAT101 deletes rows 2, 3, and 4 from C3, C4, and C5.

FIGURE 6–22

Before					After			
ROW	HT	WT	CHANGE		ROW	HT	WT	CHANGE
1	70	165	5		1	70	165	5
2	62	110	-4	→	2	66	*	*
3	64	*	*		3	68	158	-2
4	71	183	0		4	61	106	-7
5	66	*	*					
6	68	158	-2					
7	61	106	-7					

Use a colon to abbreviate a range of rows. For example:

⇨ ```
STAT> RETR 'PULSE'
STAT> DELETE 3:6 10 13 20:22 C1
```

deletes rows 3, 4, 5, 6, 10, 13, 20, 21, and 22 from column C1.

If you want to remove an entire column, use the command ERASE instead.

*Also see* COPY and ERASE.

# DESCRIBE

Calculates descriptive statistics for one or more columns.

### Syntax

| | |
|---|---|
| **DESCRIBE** | the data in C . . . C |
| **BY** | C |

### How to Use

This command prints a collection of summary statistics for each column listed:

| | |
|---|---|
| N | number of nonmissing observations. |
| NMISS | number of missing observations. |
| MEAN | average value. |
| MEDIAN | middle value. |
| TRMEAN | a 5% trimmed mean. STAT101 removes the smallest 5% and the largest 5% and computes the average value of the remaining values. |
```

STDEV	standard deviation.
SEMEAN	standard error of the mean.
MIN	smallest number.
MAX	largest number.
Q1	first quartile (25th percentile).
Q3	third quartile (75th percentile).

These summary statistics are illustrated in Fig. 6–23.

FIGURE 6–23

```
STAT> RETR 'MIAMI';
STAT> DESCRIBE 'WGT';
SUBC>    BY 'SMOKE'.

          SMOKE      N      MEAN   MEDIAN   TRMEAN   STDEV   SEMEAN
WGT         1       12    126.50   123.50   126.50   14.21    4.10
            2       71    126.66   125.00   125.98   14.90    1.77

          SMOKE    MIN      MAX       Q1       Q3
WGT         1     98.00   155.00   120.00   137.50
            2     98.00   175.00   118.00   135.00
```

You can also compute and store some of these summary statistics using the arithmetic commands.

Subcommand

BY   C

This subcommand calculates summary statistics for each distinct value in the column C, which can contain integers from −9999 to +9999 or the missing value (∗).

Also see Arithmetic and Transformations.

DIFFERENCES

Computes differences between the elements of a column.

Syntax

> **DIFFERENCES** [with lag **K**] for data in **C**, put into **C**

How to Use

This command calculates differences between elements in the input column, which usually contains time-series data. STAT101 subtracts from each row the element K rows above, and stores the differences in a new column. The first K rows of the new column will contain asterisks. If you don't specify K, the value 1 is used. See Fig. 6–24.

FIGURE 6–24

```
STAT> SET C1
DATA>    5   9   3   12   4
DATA> END
STAT> DIFF C1 C2
STAT> DIFF 2 C1 C3
STAT> PRINT C1-C3

   ROW    C1    C2    C3

     1     5     *     *
     2     9     4     *
     3     3    -6    -2
     4    12     9     3
     5     4    -8     1
```

Also see LAG.

DIR

Lists all file names in a directory.

Syntax

> **DIR**   [path]

How to Use

DIR works just as it does in the DOS operating system. To see a list of all the files in your current directory, use the command:

⇨ STAT> DIR

The following three commands would list, for example: a) all files in the diskette in drive B:, b) all files in the subdirectory C:\MARY\STATWORK, and c) all files ending with .MTW in your current directory.

⇨ STAT> DIR B:
   STAT> DIR \MARY\STATWORK
   STAT> DIR *.MTW

Make sure there is a disk inserted in your disk drive before doing a DIR of that drive. If there isn't, an error message appears, and you can put a disk in the drive and press R for retry, or F for fail if you want to cancel the DIR command. Use A for abort with care, because the computer automatically exits the STAT101 program, erasing the current worksheet.

Also see CD, TYPE, and SYSTEM.

DIVIDE

Does row-by-row division on columns and/or constants. If you divide by zero, STAT101 sets the answer to *, the missing value symbol. If you divide three columns of unequal length, STAT101 uses the shortest column as the length of the answer column.

Syntax

DIVIDE	E by E, put into E

Also see Arithmetic and Transformations.

DOTPLOT

Draws a horizontal histogram.

Syntax

DOTPLOT	of the data in C ... C
INCREMENT	= K
START	at K [end at K]
BY	C
SAME	scales for all columns

How to Use

DOTPLOT plots data from one or more columns as vertical dots above a horizontal axis. Figure 6-25 illustrates the amount of cadmium in fish livers for two groups of fish (group 2 was exposed to cadmium; group 1 was not).

FIGURE 6-25

```
STAT> RETR 'CD2'
STAT> DOTPLOT 'LIVER';
SUBC>   BY 'GROUP'.

 GROUP
 1
                   :: .
       +---------+---------+---------+---------+---------+-------LIVER
 GROUP
 2
                  .  ..   .  . :.  .          .        :
       +---------+---------+---------+---------+---------+-------LIVER
     0.000     0.080     0.160     0.240     0.320     0.400
```

Subcommands

INCREMENT   = K

This subcommand controls the distance between tick marks (+) on the horizontal axis.

| START | at K [end at K] |

Specifies the first and (optionally) the last tick mark on the horizontal axis. Any points outside this range are omitted from the display.

| BY | C |

BY creates a separate dotplot for each value in column C, which may contain integers between −9999 and +9999 or the missing value symbol (∗).

| SAME |

Tells STAT101 to use the same scale in drawing the dotplots for all columns listed.

Also see STEM-AND-LEAF, HISTOGRAM, and WIDTH.

ECHO

Causes macro commands to be displayed on the Session screen as they are executed.

Syntax

| ECHO | the commands that follow |

How to Use

This advanced command is normally used in macros (files of stored commands). When you execute a macro, the commands are printed or echoed to the Session screen as they are processed. NOECHO turns the echo off, and ECHO turns it back on again.

Also see NOECHO, EXECUTE, and STORE.

END

Tells STAT101 you are finished typing data (or macro commands).

Syntax

END

How to Use

When you are entering data from your keyboard with READ, SET, or INSERT, the (optional) END command should follow your last line of new data, as shown in Fig. 6–26. Note that the prompt at the beginning of the line changes from DATA> back to STAT> to show that STAT101 is ready for your next command.

FIGURE 6–26

```
STAT> READ C1 C2 C3
DATA>    10 -2.5  34
DATA>     5  0.5   *
DATA>     8  1.2  -6
DATA> END
      3 ROWS READ
STAT>
```

END must be used after STORE to tell STAT101 when you have finished entering the commands to be stored in a macro file.

Also see READ, SET, INSERT, and STORE.

ERASE

Removes columns and constants from the current worksheet.

Syntax

ERASE   E ... E

Use ERASE to remove any combination of columns and/or constants, including column names, from the current worksheet. (ERASE has no effect on columns or constants in saved worksheet files.) For example:

⇨ STAT> RETRIEVE 'MIAMI'
    STAT> ERASE C1-C3 'GLASSES'

erases C1, C2, and C3, along with their names, and the column named GLASSES.

Also see DELETE.

EXECUTE

Executes commands stored in a command file (macro).

Syntax

EXECUTE	commands [in **'FILENAME'**] [K times]

How to Use

STAT101 is usually used interactively: you type in a command and STAT101 immediately carries it out. There may be times, however, when you want to process several commands in a batch. To do this, use an editor or the STAT101 STORE command to put STAT101 commands in a file, often referred to as a macro. You can then use EXECUTE to carry out the commands in the macro. Online help (HELP EXECUTE) has extensive information about macros, with a number of examples. If your macro's file extension is other than the default MTB, you must include it in the 'FILENAME'.

Also see STORE, END.

EXPONENTIATE

Raises e = 2.71828 to the power E.

Syntax

> **EXPONENTIATE**   E, put into E

Also see Arithmetic and Transformations.

HEIGHT

Specifies the height of scatter plots.

Syntax

> **HEIGHT**   of plots that follow is **K** lines

How to Use

This command controls the height of PLOT, MPLOT, LPLOT, TPLOT, and TSPLOT. It does not affect the height of HISTOGRAM, DOTPLOT, STEM-AND-LEAF, or BOXPLOT. You can use any height from 5 to 400 lines; however, scales tend to be better if K is an odd number. The default height of all plots is 17 lines. For example, the command:

⇨ `STAT> HEIGHT 25`

will make subsequent plots 25 lines high until you issue another HEIGHT command.

Also see WIDTH.

HELP

Provides onscreen explanation of commands and subcommands, as well as general information about using STAT101.

Syntax

> **HELP**   [command [subcommand]]

How to Use

You may type HELP at any time during your STAT101 session—when typing commands, subcommands, or even data. If you want information about using a specific command or subcommand, use the following command; for example:

⇨ STAT> HELP HISTOGRAM BY

To see more about how HELP works, use:

⇨ STAT> HELP

From the Data Editor screen, press F1 for help.

HISTOGRAM

Draws a histogram display of a column of data.

Syntax

HISTOGRAM	of the data in C . . . C
INCREMENT	= K
START	at K [end at K]
BY	C
SAME	scales for all columns

How to Use

For each column listed, STAT101 divides the data into intervals and plots points that fall within the interval or on its lower boundary with asterisks (*). The midpoint of the interval and the count of values are also printed on each line, as shown in Fig. 6–27 on the following page.

FIGURE 6-27

```
STAT> RETR 'UTILITY'
STAT> HIST 'KWH'

Histogram of KWH    N = 44

Midpoint   Count
     400     2    **
     800     7    *******
    1200    15    ***************
    1600     8    ********
    2000     5    *****
    2400     3    ***
    2800     3    ***
    3200     1    *
```

Subcommands

INCREMENT = K

This subcommand allows you to choose the width K of each interval (i.e., the difference between midpoints).

START   at K [end at K]

START specifies the midpoint for the first and, optionally, the last interval. Any observations beyond these intervals are omitted from the display.

BY   C

BY creates a separate histogram for each distinct value in C that may contain any integers from −9999 to +9999 or the missing value symbol (∗).

To see electricity usage for each year, for example, you could use the command:

⇨ ```
STAT> HISTOGRAM 'KWH';
SUBC>   BY 'YEAR'.
```

STAT101 produces one histogram for each year.
```

SAME

This tells STAT101 to use a common scale for all histograms.

Also see STEM-AND-LEAF and DOTPLOT.

INFORMATION

Summarizes the contents of the current worksheet.

Syntax

INFORMATION   [on C . . . C]

How to Use

This command lists all the columns and stored constants contained in the current worksheet, including the column number, column name (if there is one), number of rows, and number of missing values (if any). See Fig. 6–28 for an example.

FIGURE 6–28

```
STAT> RETR 'UTILITY'
STAT> INFO

COLUMN      NAME        COUNT       MISSING
C1          MONTH       44
C2          YEAR        44
C3          KWH         44
C4          KWH $       44
C5          GAL         44          2
C6          GAL $       44          2

CONSTANTS USED: NONE
```

If you type INFO followed by a list of columns, you receive information only on those columns.

Also see DESCRIBE.

INSERT

Adds new data to columns in the worksheet.

Syntax

INSERT	data [from **'FILENAME'**]
	[between rows **K** and **K**] in **C** ... **C**
FORMAT	(format statement)
NOBS	= **K**

How to Use

With INSERT you can add new rows of data, from either the keyboard or a data file, to nonempty columns in the worksheet. INSERT does not work on empty columns or with saved worksheet files. If you don't list any row numbers, STAT101 adds the new values to the bottom of the columns. Otherwise they are inserted between the two consecutive row numbers that you specify. To insert data at the top of columns, use 0 and 1 as the row numbers.

Figure 6–29 illustrates the insertion of numbers from the keyboard.

FIGURE 6–29

```
STAT> RETRIEVE 'EXAMPLE'
STAT> PRINT C3-C4
STAT> INSERT 2 3 C3-C4
DATA>    62 105
DATA>    63 120
DATA> END
STAT> PRINT C3-C4
```

Before				After		
ROW	HT	WT		ROW	HT	WT
1	70	165		1	70	165
2	62	110		2	62	110
3	64	*		3	62	105
4	71	183	→	4	63	120
5	66	*		5	64	*
6	68	158		6	71	183
7	61	106		7	66	*
				8	68	158
				9	61	106

To insert data from the file SAMPLE.DAT between rows four and five of columns C11 and C12, for example, you could use the command:

⇨ STAT> INSERT 'SAMPLE' 4 5 C11 C12

It is not necessary to type the file extension .DAT as part of the filename.

Subcommands

FORMAT   (format statement)

Use this subcommand when you want to insert alpha data, or when you must give STAT101 special instructions (i.e., when you want to include only part of a data file).

For example, suppose you want to append the two middle columns from a file named WIDEFILE.DAT to C11 and C12 of your current worksheet. You might use the command:

⇨ STAT> INSERT 'WIDEFILE' C11 C12;
   SUBC>   FORMAT (20X,F5.2,2X,F6.0).

See online HELP INSERT FORMAT for information on using format statements with INSERT.

NOBS   K

This subcommand tells STAT101 the number of rows to be inserted.

Also see READ and SET.

INVCDF

Computes the inverse of the cumulative distribution function (CDF).

Syntax

INVCDF	for values in **E** [put into e]
BINOMIAL	n = **K**, p = **K**
CHISQUARE	degrees of freedom = **K**
DISCRETE	values in **C**, probabilities in **C**
F	df numerator = **K**, df denominator = **K**
INTEGER	discrete uniform distribution on **K** to **K**
NORMAL	[mean = **K** [standard deviation = **K**]]
POISSON	mean = **K**
T	degrees of freedom = **K**
UNIFORM	continuous distribution [on **K** to **K**]

How to Use

The CDF command computes the probability that a random variable is less than or equal to some value E. The INVCDF command performs the inverse calculation. It computes the value of E that is associated with a given cumulative probability. Figure 6–30 illustrates this relationship:

FIGURE 6–30

```
STAT> CDF 5;
SUBC>   CHISQUARE 7.

    5.0000    0.3401

STAT> INVCDF .3401;
SUBC>   CHISQUARE 7.

    0.3401    5.0005
```

(Small differences may occur due to rounding.)

To store the INVCDF, specify an additional column or constant as the command line.

If no subcommand is used, INVCDF assumes that the cumulative probabilities are from a normal distribution with a mean of 0 and a standard deviation of 1.

It is convenient to use INVCDF to look up critical values for t, F, and other distributions rather than searching through tables. For example, to find the critical, one-tailed t for an alpha of .05 and 45 degrees of freedom, use the command:

⇨ STAT> INVCDF 0.95;
    SUBC>   T 45.

         0.9500   1.6794

For the continuous distributions (CHISQUARE, F, NORMAL, T, and UNIFORM), the INVCDF exists for all probabilities between 0 and 1. (Whether or not INVCDF exists for 0 or 1 depends on the distribution.)

For the discrete distributions (BINOMIAL, DISCRETE, INTEGER, and POISSON), there may be no value that corresponds exactly to a given probability. For example, in the command

⇨ STAT> INVCDF 0.5;
    SUBC>   BINOMIAL 5 0.4.

         K P(X LESS OR = K)   K P(X LESS OR = K)
         1   0.3370   2   0.6826

the INVCDF is between 1 and 2. If the result is to be stored, the larger of the two nearest values (in our example, 2) is kept.

Subcommands

See the listing under the command RANDOM for details.

Also see RANDOM, PDF, and CDF.

IW

Sets the width for input from files.

Syntax

| IW | input width = **K** spaces |

How to Use

IW tells STAT101 how many characters per line to read from a data file. You may use any value between 9 and 160; the default is 160.

This command is useful if you want to read only the first part of an input line. For example, suppose the file HEALTH.DAT has 80 characters in each row. To read only the first 30 spaces into columns C1 to C5 of the worksheet, you would use the command:

⇨   STAT> IW 30
       STAT> READ 'HEALTH' C1-C5

Also see OW and OH.

KRUSKAL-WALLIS

This is a nonparametric test for differences among several population medians.

Syntax

KRUSKAL-WALLIS   test for data in **C**, groups in **C**

How to Use

This command extends the two-sample Mann-Whitney test to any number of groups, and represents a nonparametric alternative to one-way analysis of variance. It computes the test statistic H, which is used to test the null hypothesis regarding the differences among the population medians. The first column in Fig. 6–31, 'GAL $', contains sample data for all the populations, which are identified by group numbers in the second column, 'YEAR'.

FIGURE 6-31

```
STAT> RETR 'UTILITY'
STAT> KRUS 'GAL $' 'YEAR'

  42 CASES WERE USED
   2 CASES CONTAINED MISSING VALUES

LEVEL    NOBS    MEDIAN   AVE. RANK   Z VALUE
 1983     12     12.06      19.8      -0.56
 1984     12     12.65      20.5      -0.33
 1985     12     12.71      23.0       0.49
 1986      6     13.35      23.9       0.52
OVERALL   42                21.5

H = 0.70   d.f. = 3   p = 0.872
H = 0.70   d.f. = 3   p = 0.872  (adj. for ties)
```

For each group or level, STAT101 prints the number of observations: the median, the average rank, and the z-value (which indicates how the average rank for that group differs from the average rank for all groups).

STAT101 also prints a modified H, adjusted for ties. H and H(ADJ. FOR TIES) are equal if there are no ties.

Finally, STAT101 prints a message if one or more groups are small (less than 5 observations).

Also see **MANN-WHITNEY, AOVONEWAY,** and **ONEWAY.**

LAG

Moves numbers in a column down a specified number of rows.

Syntax

LAG   [of length **K**] data in **C**, put into **C**

How to Use

This time-series command moves the row elements of a column down K rows, with 1 as the default value of K, and stores the result in a new column of the same length. There will be K missing value symbols (*) at the top of the output column, while the last K values from the input column will be dropped, as shown in Fig. 6-32 on the following page.

FIGURE 6-32

```
STAT> SET C1
DATA>   1 12 3 9 1
DATA> END
STAT> LAG 2 C1 C2
STAT> PRINT C1 C2

ROW    C1    C2
 1      1     *
 2     12     *
 3      3     1
 4      9    12
 5      1     3
```

You can also copy a column with a lag of 1, using a LET expression; for example:

⇨ STAT> LET C3 = LAG(C1)

Also see LET, ACF, PACF, and CCF.

LET

Changes the value of a number in a column, or does arithmetic, using an algebraic expression.

Syntax

LET   E = expression

Also see Arithmetic and Transformations.

LOGE

Calculates logarithms to the base e (natural logs). If you take the log of zero or a negative number, STAT101 sets the result to *, missing value, and prints an "out of bounds" message.

Syntax

LOGE (natural logarithm) of **E**, put into **E**

Also see Arithmetic and Transformations.

LOGTEN

Calculates logarithms to the base 10. If you take the log of zero or a negative number, STAT101 sets the result to *, missing value, and prints an "out of bounds" message.

Syntax

LOGTEN (logarithm-base 10) of **E**, put into **E**

Also see Arithmetic and Transformations.

LPLOT

Draws a labeled plot, using group codes as the plotting symbols.

Syntax

LPLOT   C vs C, groups in C

  YINCREMENT   = K

  YSTART   at K [end at K]

  XINCREMENT   = K

  XSTART   at K [end at K]

How to Use

This command plots the first column (the y-axis) against the second column (the x-axis), marking points with letters corresponding to the group numbers in the third column. All three columns must be the same length. An (x,y) pair with 1 as its group number is plotted with an A, and so on; a point with an * in the group column is plotted with an *. Figure 6–33 displays the complete conversion chart of group numbers to letters.

FIGURE 6–33

−2	X
−1	Y
0	Z
1	A
2	B
3	C
4	D
5	E
6	F
7	G
8	H
9	I
10	J
11	K
12	L
13	M
14	N
15	O
16	P

17	Q
18	R
19	S
20	T
21	U
22	V
23	W
24	X
25	Y
26	Z
27	A
28	B
29	C
and so on	

The group column can be coded with this chart in mind. For example, to plot data with M's for males and F's for females, you could use the codes 13 and 6 in the column specifying sex. (This is easy to change with the CODE command.)

In Fig. 6–34, electricity-use data for each month are plotted with a different letter. As you would expect, the highest electricity use occurs in the months January and February, which correspond to the letters A and B in the plot.

FIGURE 6-34

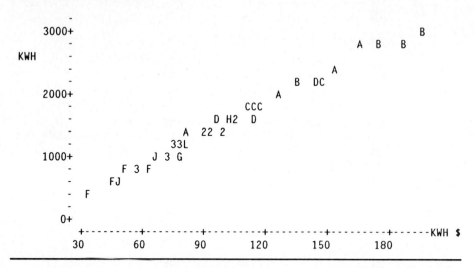

If several points fall on the same spot, a count is given, as in the PLOT command.

Use the commands HEIGHT and WIDTH to control the size of the LPLOT if desired. You may override STAT101's default scaling with the subcommands listed above, which are described further under the command PLOT.

Also see PLOT, MPLOT, TPLOT, CODE, WIDTH, and HEIGHT.

MANN-WHITNEY

Performs a two-sample rank test.

Syntax

MANN-WHITNEY [alternative K] [K % confidence] for the data in C and C

How to Use

This nonparametric command performs a two-sample rank test for the difference between two population medians, and calculates corresponding point and confidence interval estimates. It is sometimes called the two-sample Wilcoxon rank sum test.

STAT101 calculates an approximate 95% confidence interval by default, and computes the attained significance level (p-value). For a one-sided test, enter 1 as the first value of K to test the alternative hypothesis H_1:(median1 > median2), or −1 for the alternative hypothesis H_1:(median1 < median2). Specify a second value of K to change the default confidence level. STAT101 will find a confidence interval for the difference between median 1 and median 2 that comes as close to the specified confidence as possible. See Fig. 6–35 for an example.

FIGURE 6–35

```
STAT> SET C1
DATA>    90 72 61 66 81 69 59 70
DATA> END
STAT> SET C2
DATA>    62 85 78 66 80 91 69 77 84
DATA> END
STAT> MANN-WHITNEY C1 C2

Mann-Whitney Confidence Interval and Test

MONTH      N =    8    Median =        69.50
YEAR       N =    9    Median =        78.00
Point estimate for ETA1-ETA2 is         -7.50
95.1 pct c.i. for ETA1-ETA2 is (-18.00,4.00)
W = 60.0
Test of ETA1 = ETA2 vs.  ETA1 n.e. ETA2 is significant at 0.2685
The test is significant at 0.2679 (adjusted for ties)

Cannot reject at alpha = 0.05
```

Also see TWOSAMPLE and WTEST.

MAXIMUM

Syntax

MAXIMUM	largest number in C [put in K]

Also see Arithmetic and Transformations.

MEAN

Syntax

MEAN	average all values in **C** [put in **K**]

Also see Arithmetic and Transformations.

MEDIAN

Syntax

MEDIAN	middle data value in **C** [put in **K**]

Also see Arithmetic and Transformations.

MINIMUM

Syntax

MINIMUM	smallest number in **C** [put in **K**]

Also see Arithmetic and Transformations.

MPLOT

Plots multiple pairs of columns on the same axes.

Syntax

MPLOT	**C** vs **C** and **C** vs **C** and . . . **C** vs **C**
YINCREMENT	= **K**

YSTART	at K [end at K]
XINCREMENT	= K
XSTART	at K [end at K]

How to Use

This command plots several pairs of (x,y) coordinates, representing two columns each, on the same set of axes. The first pair of columns is plotted with the symbol A, the second with B, etc. If several points fall on the same spot, a count is given, as in PLOT.

Figure 6–36 illustrates a regression that shows how blood pressure ('BP', number of points above normal blood pressure) varies with weight ('OVER WGT', number of kilograms above normal weight). In Fig. 6–36, STAT101 stores the standardized residuals and fits (y-hats) in C5 and C6, respectively. Use MPLOT to see the original data plotted with A's, and the fitted regression line, plotted with B's, on the same plot.

FIGURE 6–36

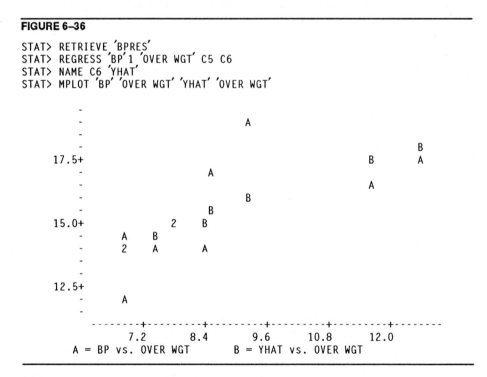

Use the commands HEIGHT and WIDTH to control size of the MPLOT if desired. You may override STAT101's default scaling of MPLOT with the subcommands listed above, which are described further under the command PLOT.

Also see PLOT, LPLOT, TPLOT, WIDTH, and HEIGHT.

MULTIPLY

Does row-by-row multiplication on columns and/or constants. You may use up to 50 arguments (49 multipliers plus one storage column). If you multiply columns of unequal length, STAT101 uses the shortest column as the length of the answer column.

Syntax

MULTIPLY   E by E . . . by E, put into E

Also see Arithmetic and Transformations.

N

Counts the number of nonmissing values in the column.

Syntax

N   number of nonmissing values in C [put in K]

Also see Arithmetic and Transformations.

NAME

Names columns. You can also name columns in the Data Editor; see Chapter 3, Using the Data Editor.

Syntax

NAME	for C is 'NAME' . . . for C is 'NAME'

How to Use

Use this command to assign a name, up to eight characters long, to one or more columns. You can then refer to the column by its name or number in any STAT101 command. Column names must always be enclosed in single quotation marks. For example, you could use the command:

⇨ STAT> NAME C5 'Volume' C6 'Diameter'

Column names must not begin or end with a space, contain the symbols # or ', or be the same as the saved-worksheet name or another column name; all other characters are valid. The command INFO shows you a list of column names already assigned. You can change the name of a column at any time with another NAME command. (The command ERASE removes both data and the column name.) To remove a column name without changing data in the column, use the command:

⇨ STAT> NAME C5 ''

C5 is followed by two single quotation marks, not a double quotation mark.

Also see INFORMATION.

NEWPAGE

Goes to the top of the next page.

Syntax

NEWPAGE

How to Use

Use this command when either the PAPER or OUTFILE command is in effect to tell STAT101 that the next command you type should start at the top of a new page. (Remember that PAPER sends a copy of your STAT101 session to a printer; OUTFILE sends it to a file.) The word NEWPAGE will not be printed.

Also see OUTFILE, PAPER, and OH.

NMISS

Counts the number of missing values in the column.

Syntax

| NMISS | number of missing values in **C** [put in **K**] |

Also see Arithmetic and Transformations.

NOCONSTANT

Fits models without a constant term in all REGRESS and STEPWISE commands.

Syntax

| NOCONSTANT |

How to Use

STAT101 normally fits a constant term (the y-intercept of the regression line) to all REGRESS and STEPWISE models. The command NOCONSTANT causes STAT101 to omit the constant term from subsequent REGRESS and STEPWISE models until you issue a CONSTANT command. When NOCONSTANT is in effect, REGRESS and STEPWISE output excludes the R-squared value since its interpretation is difficult.

Also see CONSTANT, REGRESS, and STEPWISE.

NOECHO

Suppresses the printing of commands on the screen during macro execution.

Syntax

NOECHO   the commands that follow

How to Use

This advanced command, along with ECHO, controls whether or not commands are printed or echoed on the screen as a command file (macro) is processed.

Also see ECHO and EXECUTE.

NOOUTFILE

Stops sending output to a file.

Syntax

NOOUTFILE

How to Use

The OUTFILE command causes practically everything that appears on your screen, commands and output, to be saved in a specified output file. NOOUTFILE tells STAT101 to stop sending output to the file. No file name is required with the NOOUTFILE command.

Also see OUTFILE, PAPER, and NOPAPER.

NOPAPER

Stops sending output to the printer.

Syntax

NOPAPER

How to Use

The PAPER command causes practically everything you see on your screen, commands and output, to go directly to a printer as well as the screen. NOPAPER tells STAT101 to stop printing and send output to the screen only.

Also see PAPER, OUTFILE, and NOOUTFILE.

NOTE

Precedes comments in a STAT101 macro.

Syntax

NOTE   [any text]

How to Use

Like the comment symbol #, NOTE is used to enter comments. For example, a line in a command file (macro) might be:

⇨   NOTE The table below summarizes group means.

Unlike #, NOTE is a command and so must be the first word on a line. Moreover, NOTE commands are printed or echoed in a special way in macros.

Also see #, NOECHO, ECHO, and EXECUTE.

NSCORE

Calculates normal scores that are used to do normal probability plots and normal scores tests.

Syntax

NSCORE   (normal scores) of C, put into C

Also see Arithmetic and Transformations.

OH

Controls height of output.

Syntax

OH   output height = **K** lines

How to Use

This command sets the output height, the maximum number of lines displayed on your terminal screen before the Continue? prompt appears. The default value of OH is 24; you may specify any nonnegative integer. Setting OH to zero suppresses the Continue? prompt entirely, so that output scrolls continuously.

As a subcommand to PAPER or OUTFILE, OH controls the page length of the file going to your printer or the outfile. By default, there are 66 lines per page in an output file.

Also see OW, IW, OUTFILE, and PAPER.

ONEWAY

Computes a one-way analysis of variance.

Syntax

ONEWAY   on the data in C, subscripts in C [residuals in C [fitted values in C]]

How to Use

This command performs one-way analysis of variance, as does AOVONEWAY. While AOVONEWAY requires that data for each factor be in a separate column, ONEWAY expects data for all factors to be in the first column listed, with codes identifying the factor or level number in the second column. Codes may be any integers from −9999 to +9999 or the missing value symbol (∗).

You can store residual values if you list a third column on the command line, fitted values if you list a fourth column. Figure 6–37 compares the cost of electricity over several years.

FIGURE 6–37

```
STAT> RETR 'UTILITY'
STAT> ONEWAY 'KWH $' 'YEAR' C10 C11

ANALYSIS OF VARIANCE ON KWH $
SOURCE       DF         SS        MS         F       p
YEAR          3       2530       843      0.56   0.642
ERROR        40      59737      1493
TOTAL        43      62267
                                        INDIVIDUAL 95 PCT CI'S FOR MEAN
                                        BASED ON POOLED STDEV
LEVEL         N       MEAN     STDEV     ---------+---------+---------+-------
 1983        12      85.50     23.73     (-----------*-----------)
 1984        12     101.17     43.11                 (-----------*-----------)
 1985        12      96.24     42.92           (-----------*-----------)
 1986         8     106.60     42.82                   (-------------*-------------)
                                        ---------+---------+---------+-------
POOLED STDEV =       38.64                      80        100       120
```

STAT101 stores residual values in column C10, fitted values in C11.

Also see AOVONEWAY, TWOT, and KRUSKAL-WALLIS.

OUTFILE

Stores a copy of your STAT101 session in a file.

Syntax

OUTFILE	'FILENAME'
OH	output height is K lines
OW	output width is K spaces
NOTERMINAL	

How to Use

After you type the OUTFILE command, the commands and output shown on your screen are saved in the specified file until you type NOOUTFILE or end your STAT101 session. (OUTFILE and PAPER cannot both be in effect simultaneously. Thus, OUTFILE can also be canceled by typing the PAPER command.) To include your data in the file, use the PRINT command while OUTFILE is still in effect. Be sure to enclose the file name in single quotation marks.

For example, Fig. 6–38 shows how to create an outfile called STATWORK, in which you retrieve the data set WGTHGT (weights and heights of first graders) and then plot a histogram of C1 'WEIGHT':

FIGURE 6–38

```
STAT> OUTFILE 'STATWORK'
STAT> RETR 'WGTHGT'
STAT> HIST C1
STAT> NOOUTFILE
```

Figure 6–38 only shows the commands. If you wanted to see your outfile, you could use the TYPE command:

⇨ `STAT> TYPE STATWORK.LIS`

STAT101 supplies an LIS file extension to the name automatically. After you exit from STAT101, you can view and edit this file with any word processing program or print it out. You may need to check with your instructor about where you should send your outfile (that is, to a diskette or to a hard disk). If you are sending your outfile to a disk on the A: drive, replace the command OUTFILE 'STATWORK' with OUTFILE 'A:\STATWORK'.

If you use the same file name with OUTFILE again later, STAT101 appends new output to the original file. While OUTFILE is in effect, typing NEWPAGE will cause the printer to skip to the top of a new page when you print the output file.

Subcommands

OH	output height is K lines

The default value of OH is 66 lines per page in an outfile. Use the OH subcommand to change output height, if desired. An output height of zero suppresses all paging in the outfile.

OW   output width if K spaces

Output files normally contain 79 spaces per line. To change the line width, enter any number from 30 to 132 with the OW subcommand. Note that some commands cannot produce output narrower than 70 spaces.

NOTERMINAL

This subcommand instructs STAT101 to stop printing output on your terminal screen, but to continue sending results to the outfile.

Also see NOOUTFILE, PAPER, NOPAPER, OH, OW, and NEWPAGE.

OW

Sets the line width for output.

Syntax

OW   output width = K spaces

How to Use

This command sets the maximum width for each line of output on your terminal screen. The default output width is 79 spaces; you may choose any width from 30 to 132. Note that certain commands cannot produce output narrower than 70 spaces, regardless of the OW setting.

OW may also be used as a subcommand with OUTFILE or PAPER to set the maximum line width for an outfile or printed output.

Also see OH, IW, OUTFILE, and PAPER.

PACF

Displays partial autocorrelations for time-series data in a graph.

Syntax

PACF	[up to **K** lags] for data in **C** [put into **C**]

How to Use

This command computes, plots, and optionally stores partial autocorrelations for a time series, with time lags varying from 1 to K. If you don't specify K, STAT101 uses the value $\sqrt{n} + 10$, where n = number of observations. In Fig. 6–39, partial autocorrelations are computed and stored in column C10

FIGURE 6–39

```
STAT> RETR 'UTILITY'
STAT> PACF 3 'KWH $' C10

PACF of KWH $
         -1.0 -0.8 -0.6 -0.4 -0.2  0.0  0.2  0.4  0.6  0.8  1.0
           +----+----+----+----+----+----+----+----+----+----+
    1  0.530                              XXXXXXXXXXXXX
    2 -0.238                       XXXXXXX
    3 -0.182                        XXXXXX
```

Also see ACF, CCF, and LAG.

PAPER

Sends a copy of your STAT101 session to the printer.

Syntax

PAPER		
	OH	output height is **K** lines
	OW	output width is **K** spaces
	NOTERMINAL	

How to Use

After you type this command, practically all the commands and output on your screen are sent to the printer until you type NOPAPER or end your STAT101 session. PAPER and OUTFILE cannot both be in effect simultaneously. Thus, PAPER can also be cancelled by typing the OUTFILE command.

In order for PAPER to work, you must have a printer connected to your computer. If you are working on a network, where more than one computer could be linked to a printer, do not use the PAPER command without your instructor's or lab director's explicit permission. If you're not sure you have the proper setup, use OUTFILE to save a copy of your session in a file that you can print out after you exit STAT101.

While PAPER is in effect, typing NEWPAGE will cause the printer to skip to the top of a new page.

Subcommands

OH   output height is **K** lines

There are normally 66 lines of output per page. Use the OH subcommand to change output height if desired. An output height of zero suppresses all paging.

OW   output width is **K** spaces

Printed output is normally 79 spaces wide. To change the output width, enter any number from 30 to 132 with the OW subcommand. Some commands cannot produce output less than 70 spaces wide regardless of the OW setting.

NOTERMINAL

This subcommand tells STAT101 to stop displaying output on the screen, but to continue sending it to the printer.

Also see NOPAPER, OUTFILE, OH, OW, and NEWPAGE.

PARPRODUCTS

Calculates partial products. The product of the first i rows of the first column is put into the i-th row of the second column.

Syntax

PARPRODUCTS (partial products) of **C**, put into **C**

Also see Arithmetic and Transformations.

PARSUMS

Calculates partial sums. The sum of the first i rows of the first column is put into the i-th row of the second column.

Syntax

PARSUMS (partial sums) of **C**, put into **C**

Also see Arithmetic and Transformations.

PDF

Computes probabilities, or the probability density function, for the given values from a specified distribution.

Syntax

PDF	for values in **E** [put into **E**]
BINOMIAL	n = **K**, p = **K**
CHISQUARE	degrees of freedom = **K**
DISCRETE	values in **C** probabilities in **C**

(continued)

F	df numerator = **K**, df denominator = **K**
INTEGER	discrete uniform distribution on **K** to **K**
NORMAL	[mean = **K** [standard deviation = **K**]]
POISSON	mean = **K**
T	degrees of freedom = **K**
UNIFORM	continuous distribution [on **K** to **K**]

How to Use

This command computes the probability or the probability density function associated with a constant or column of numbers. The distribution can be specified with any of the subcommands listed above; otherwise STAT101 assumes a normal distribution with a mean of 0 and standard deviation of 1.

For continuous distributions (CHISQUARE, F, NORMAL, T, and UNIFORM), PDF calculates the continuous probability density function, the height of the curve for each number specified. For example, you could compute the PDF for the numbers 28 through 32 from a normal distribution with a mean of 35 and a standard deviation of 2, as shown in Fig. 6–40.

FIGURE 6–40

```
STAT> SET C4
DATA>   28:32
DATA> END
STAT> PDF C4 C5;
SUBC>   NORMAL 35 2.
STAT> PRINT C4 C5

 ROW     C4          C5
   1     28      0.0004363
   2     29      0.0022159
   3     30      0.0087642
   4     31      0.0269955
   5     32      0.0647588
```

Since a storage location (column C5) was given for the PDF values, they were not displayed on the screen. They are displayed automatically if no storage is specified.

For discrete distributions (BINOMIAL, POISSON, INTEGER, and DISCRETE), PDF calculates the probability of occurrence of the values indicated on the command line. You may also omit the storage argument to have STAT101 produce a full table of probabilities for the discrete distribution chosen.

Subcommands

See the listing under the command CDF for details.

Also see RANDOM, CDF, and INVCDF.

PLOT

Plots one column against another.

Syntax

PLOT	C versus C	
YINCREMENT	= K	
YSTART	at K [end at K]	
XINCREMENT	= K	
XSTART	at K [end at K]	

How to Use

This command plots the data in the first column on the (vertical) y-axis versus the data in the second column on the (horizontal) x-axis, as shown in Fig. 6–41 on the following page.

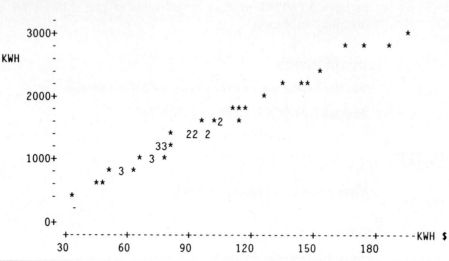

FIGURE 6-41

```
STAT> RETRIEVE 'UTILITY'
STAT> PLOT 'KWH' 'KWH $'
```

Individual points are plotted with an *. When two or more points coincide, STAT101 prints a count instead. If more than 9 points fall on the same spot, the symbol + is printed.

Use the commands HEIGHT and WIDTH to control the size of the PLOT if desired. You may override STAT101's default scaling with the subcommands described below.

Subcommands

YINCREMENT = K

YINCREMENT controls the vertical scale: K is the distance between the tick marks (+) on the y-axis of the plot.

YSTART   at K [end at K]

This subcommand specifies the first and, optionally, the last value to be represented on the vertical axis. Data points outside this range are excluded.

XINCREMENT = K

XINCREMENT controls the horizontal scale: K is the distance between the tick marks (+) on the x-axis.

XSTART at K [end at K]

This subcommand specifies the first and, optionally, the last value to be represented on the horizontal axis. Data points outside this range are omitted.

Also see LPLOT, MPLOT, WIDTH, and HEIGHT.

PRINT

Displays the data in the current worksheet on the screen.

Syntax

PRINT view the data in E . . . E

FORMAT (format statement)

How to Use

PRINT displays data from the worksheet on your screen. You may list any combination of columns or constants in any order with PRINT. See Fig. 6–42.

FIGURE 6–42

```
STAT> RETRIEVE 'EXAMPLE'
STAT> PRINT K1-K3 C1-C4
K1        10.0000
K2        25.0000
K3         *
```

ROW	SEX	DIET	HT	WT
1	1	NO	70	165
2	2	YES	62	110
3	1	YES	64	*
4	1	YES	71	183
5	2	NO	66	*
6	1	NO	68	158
7	2	YES	61	106

If only one column is listed, it appears across the screen. Multiple columns are printed vertically. STAT101 chooses the number of decimal digits for each number unless you give it specific instructions with the FORMAT subcommand. You can also change the width of the display with the OW command.

PRINT does not send data to your printer. One way to get a paper copy of data in your worksheet is to use PRINT while an OUTFILE or PAPER command is in effect. After you exit from STAT101 you can edit the file, if you wish, and then print it.

Subcommand

FORMAT

This subcommand tells STAT101 exactly how and where to print column data on your screen. You may not use a format to print stored constants. See online help, HELP PRINT FORMAT, for more information.

Also see WRITE, READ, and OUTFILE.

RAISE

Raises each value to a specified power, working row by row on columns and/or constants. The power may be any real number. If STAT101 cannot do the calculation, like -1 to the power .5, then it sets the answer to *, missing value.

Syntax

RAISE	E to the power of E, put into E

Also see Arithmetic and Transformations.

RANDOM

Generates random samples of numbers.

Syntax

RANDOM	K observations into each C . . . C
BERNOULLI	p = K
BINOMIAL	n = K, p = K
CHISQUARE	degrees of freedom = K
DISCRETE	values in C probabilities in C
F	df numerator = K, df denominator = K
INTEGER	discrete uniform distribution on K to K
NORMAL	[mean = K [standard deviation = K]]
POISSON	mean = K
T	degrees of freedom = K
UNIFORM	continuous distribution [on K to K]

How to Use

With this command, you can generate a random sample of K values from a given distribution in each column listed. If you omit a subcommand, STAT101 assumes a normal distribution with a mean of zero and a standard deviation of one. Unless you use the BASE command, RANDOM will generate different numbers each time you use it.

The example in Fig. 6–43 generates five values into columns C1 and C2 from a normal distribution with a mean of 100 and a standard deviation of 15. It also generates six integers between 5 and 25 into columns C3 to C5.

FIGURE 6-43

```
STAT> RANDOM 5 C1 C2;
SUBC>    NORMAL 100 15.
STAT> RANDOM 6 C3-C5;
SUBC>    INTEGER 5 25.
STAT> PRINT C1-C5

  ROW       C1         C2       C3    C4    C5
    1    129.110    91.042     11    19    20
    2    105.784    89.685      6    12    25
    3     91.429    94.483     20     9    25
    4    103.411   104.968      8    12    21
    5    106.818   103.177     13    11    16
    6                          13    11     7
```

Your data will of course be different, as STAT101 will generate different random values.

Subcommands

BERNOULLI p = K

Each Bernoulli trial results in success (1) or failure (0), with p as the probability of success.

BINOMIAL n = K, p = K

The binomial is the distribution of the number of successes after n Bernoulli trials, with p as the probability of success on each trial.

CHISQUARE degrees of freedom = K

Specify the degrees of freedom for the chisquare distribution.

DISCRETE values in C, probabilities in C

You can describe your own discrete distribution. First, enter the list of discrete values and their associated probabilities into two worksheet columns. Then use the DISCRETE subcommand, specifying those two columns.

For example, five random values can be generated into C7, based on the discrete distribution in C5 and C6, as shown in Fig. 6-44.

FIGURE 6-44

```
STAT> READ C5 C6
DATA>   -1    0.25
DATA>    0    0.50
DATA>    1    0.25
DATA> END
STAT> RANDOM 5 C7;
SUBC>    DISCRETE C5 C6.
STAT> PRINT C7
C7
     0    1   -1    0    1
```

Your random data will of course be different.

F   df numerator = K, df denominator = K

For the F-distribution, you must specify the associated degrees of freedom for the numerator and denominator.

INTEGER   discrete uniform distribution on K to K

In this distribution, each integer value in the interval from K to K, inclusive, occurs with equal probability. K may be any integer from −32767 to +32767.

NORMAL   [mean = K [standard deviation = K]]

The normal distribution is determined by its mean and its standard deviation. If the K's are omitted, the mean is assumed to be 0 and the standard deviation is assumed to be 1.

POISSON   mean = K

The specified mean must be greater than 0 and less than or equal to 50.

> **T** degrees of freedom = K

This subcommand specifies the Student's t distribution with K degrees of freedom.

> **UNIFORM**   continuous distribution [on K to K]

This distribution covers the interval from K to K. If the Ks are omitted, then the interval from 0 to 1 is used.

Also see PDF, CDF, INVCDF, BASE, and SAMPLE.

RANK

Assigns rank scores to values in a column.

Syntax

> **RANK**   the data in **C**, put ranks in **C**

How to Use

RANK assigns the number 1 to the smallest value in the first column, 2 to the next smallest value, and so on. Ties are each assigned the average of their ranks, as shown in Fig. 6–45.

FIGURE 6–45

```
STAT> SET C1
DATA>   8 3 1 3 7
DATA> END
STAT> RANK C1 C2
STAT> PRINT C1 C2

ROW    C1    C2
 1      8    5.0
 2      3    2.5
 3      1    1.0
 4      3    2.5
 5      7    4.0
```

Also see SORT, Arithmetic and Transformations.

RCOUNT

Counts the number of observations (missing and nonmissing) in each row.

Syntax

RCOUNT	total number of values per row of E . . . E, put in C

Also see Arithmetic and Transformations.

READ

Enters data row by row into worksheet columns. You can also enter data into the Data Editor; see Chapter 3.

Syntax

READ	data [from **'FILENAME'**] into C . . . C
FORMAT	(format statement)
NOBS	= K

How to Use

READ enters data one row at a time into one or more worksheet columns, either from the keyboard or from a data file. Any existing values in the columns will be erased (though column names will not be erased). (Compare this command to INSERT, which adds data to existing columns, or SET, which enters a single column of data.)

Each row of data must start on a new line and contain one value per column. At least one space or comma should separate the numbers (unless the FORMAT subcommand is used). Letters or words on the data line are normally ignored. (To read in alpha data, use the FORMAT subcommand.) The missing value code (*) and data in exponential (scientific) notation, such as 1.23E-2, are valid entries. Type END when you are finished, as shown in Fig. 6–46 on the following page.

FIGURE 6-46

```
STAT> READ C1 C2
DATA>    1     -3
DATA>    0      *
DATA>    1.5  -50
DATA> END
STAT> PRINT C1 C2

 ROW     C1    C2
  1     1.0   -3
  2     0.0    *
  3     1.5  -50
```

In Fig. 6-46, C1 now contains 1.0, 0.0, and 1.5, while C2 contains -3, *, and -50.

The command

⇨ STAT> READ 'MIAMI' C1-C24

enters 24 columns of data from the file MIAMI.DAT. Use READ with ASCII text files, such as those created with WRITE (with the extension .DAT); RETRIEVE is for saved worksheet files (with the extension .MTW).

Subcommands

FORMAT   (format statement)

Use this subcommand with a format statement using Fortran format codes when you want to enter alpha data, or when you must give STAT101 special instructions.

FIGURE 6-47

```
STAT> READ C1 C2;
SUBC>    FORMAT (A3,1X,F2.0).
DATA>    LOW 25
DATA>    MED 30
DATA>    HI  20
DATA> END
```

In Fig. 6-47, STAT101 reads alpha data into the first three spaces (C1), skips a space, reads a number into the next two spaces (C2), and then continues until there are no more data rows. Note that you must add an extra space in the third line because HI has only two letters, and the A3 code tells STAT101 to expect three alpha characters. Use the HELP READ FORMAT command for more explanation of using FORMAT with READ.

You may enter up to 80 characters (A80 format) into an alpha column with READ, with SET, and INSERT. Two or more alpha columns can be pasted together into a new column, up to 80 characters wide, with the command CONCATENATE.

NOBS

You may use this subcommand to indicate how many rows will be entered. The END command is not necessary when NOBS is used.

Also see SET, INSERT, WRITE, and CONCATENATE.

REGRESS

Fits a linear regression model to the data.

Syntax

REGRESS	y in **C** on **K** predictors in **C** . . . **C** [standardized residuals in **C** [fitted values in **C**]]
NOCONSTANT	
COEFFICIENTS	put in **C**
PREDICT	for **E** . . . **E**
RESIDUALS	put into **C**
DW	(Durbin-Watson statistic)

How to Use

REGRESS fits a linear equation to your data by the least squares method. Specify the dependent or y variable, the number of predictors, and the columns containing the predictor variables. You may specify an additional column to store the standardized residuals, and another to store the fitted values. Figure 6–48 shows a simple regression example, in which blood pressure (number of points above normal) is examined as a function of weight (number of kilograms over normal weight).

FIGURE 6-48

```
STAT> RETRIEVE 'BPRES'
STAT> REGRESS 'BP' 1 'OVER WGT' C5 C6;
SUBC>    COEFFICIENTS C7;
SUBC>    RESIDUALS C8.

The regression equation is
BP = 9.29 + 0.704 OVER WGT

Predictor         Coef        Stdev      t-ratio         p
Constant         9.291        2.737         3.40     0.012
OVER WGT        0.7041       0.3027         2.33     0.053

s = 1.755      R-sq = 43.6%      R-sq(adj) = 35.5%

Analysis of Variance

SOURCE         DF          SS          MS          F         p
Regression      1      16.678      16.678       5.41     0.053
Error           7      21.571       3.082
Total           8      38.249
```

In Fig. 6–48, the REGRESS command stores standardized residuals in C5, fitted values (y-hats) in C6, regression coefficients in C7, and residuals (observed minus fitted values) in C8. To view the contents of these columns, use the PRINT command.

STAT101 prints the regression equation, followed by a table listing the coefficient, standard deviation, t-ratio, and p-value for each variable in the model. The output also includes s, the estimated standard deviation about the regression line (equal to the square root of MS Error), R-squared (the coefficient of determination), and R-squared adjusted for degrees of freedom, followed by the analysis of variance table.

Finally, STAT101 lists unusual observations (if there are any), noted with an X if the predictors are unusual (leverage value is large) or with an R if the response is unusual (standardized residual is greater than 2). Leverage values greater than 3p/n or .99, whichever is smaller, where p is the number of predictors and n is the number of observations, are considered large.

The BRIEF command can be issued before REGRESS to increase or decrease the amount of output displayed.

Subcommands

NOCONSTANT

This subcommand tells STAT101 not to fit the intercept (where the line crosses the y-axis) term in the regression model. NOCONSTANT can also be used as a main command before the REGRESS command.

COEFFICIENTS put into C

Stores coefficients of the regression model in the specified column.

PREDICT for E . . . E

The PREDICT subcommand computes fitted values based on the regression equation. To use it, specify one argument (column or constant) for each predictor on the main command line. Several PREDICT subcommands may be used with the same REGRESS command. REGRESS output will include associated fitted values plus their 95% confidence intervals and 95% prediction intervals.

For example, using the same BPRES.MTW data set used in Fig. 6–48, you could predict blood pressure values for two sets of predictor values by typing:

⇨
```
STAT> REGRESS 'BP' 3 'OVER WGT' 'FATS' 'EXERCISE';
SUBC>    PREDICT 14.3  4.5   9.6;
SUBC>    PREDICT  8.6  6.2  10.3.
```

If you specify an unusual case (leverage value greater than $3p/n$, where p is the number of predictors and n the number of observations), STAT101 prints an X at the end of the output line. Extreme cases (leverage value greater than $5p/n$) are marked with XX.

RESIDUALS put into C

Differences between observed y-values and fitted y-values are stored in the specified column.

DW (Durbin-Watson statistic)

Prints the Durbin-Watson statistic for autocorrelation in the data.

Also see BRIEF, NOCONSTANT, and STEPWISE.

RESTART

Begins a new STAT101 session without exiting the program.

Syntax

RESTART

How to Use

RESTART is a quick way to erase all the data in the worksheet and reset all parameters (e.g., BRIEF, OH, WIDTH) to their default values without leaving STAT101. RESTART also closes all open files. Be sure to save any important data before using this command.

Also see ERASE and SAVE.

RETRIEVE

Enters a saved worksheet file into the current worksheet.

Syntax

RETRIEVE   ['FILENAME']

PORTABLE

How to Use

RETRIEVE erases all the data in the current worksheet and replaces it with data from a saved worksheet file. Be sure to save any important data before using this command. RETRIEVE can only enter a file created with the STAT101 command SAVE. All the columns, column names, and stored constants contained in the worksheet when it was saved will be present when it is retrieved.

The file name must be enclosed in single quotes and should include path information (drive letter and/or directory name) if appropriate. For example:

⇨ `STAT> RETRIEVE 'STA261'`
`WORKSHEET SAVED  3/ 3/1993`

`Worksheet retrieved from file: STA261.MTW`

Alternatively, typing

⇨ `STAT> RETRIEVE 'B:STA261'`

would enter the contents of the file STA261.MTW on a disk in drive B: into the current worksheet. The default file extension for a saved STAT101 worksheet is .MTW. If the RETRIEVE command is issued without a file name, STAT101 searches for STAT101.MTW. If the file name includes an extension different from MTW, you must include it in the 'FILENAME'.

The lines of information STAT101 prints when you RETRIEVE a file have been omitted from other examples in this chapter to save space.

Subcommand

PORTABLE

This subcommand retrieves a saved worksheet created by STAT101 on a different computer type using SAVE with the PORTABLE subcommand. This retrieval takes a few seconds longer than the retrieval of a regular saved STAT101 worksheet.

Also see SAVE.

RMAXIMUM

Finds the maximum value in each row, omitting missing observations.

Syntax

RMAXIMUM   largest number per row of E . . . E, put in C

Also see Arithmetic and Transformations.

RMEAN

Calculates the mean of the numbers in each row, omitting missing observations.

Syntax

RMEAN   average of all values per row of E . . . E, put in C

Also see Arithmetic and Transformations.

RMEDIAN

Calculates the median of the numbers in each row, omitting missing observations.

Syntax

RMEDIAN   middle data value per row of E . . . E, put in C

Also see Arithmetic and Transformations.

RMINIMUM

Finds the minimum value in each row, omitting missing values.

Syntax

RMINIMUM   smallest number per row of E . . . E, put in C

Also see Arithmetic and Transformations.

RN

Counts the number of nonmissing values in each row.

Syntax

RN   number of nonmissing values per row of **E** ... **E**, put in **C**

Also see Arithmetic and Transformations.

RNMISS

Counts the number of missing values in each row.

Syntax

RNMISS   number of missing values per row of **E** ... **E**, put in **C**

Also see Arithmetic and Transformations.

ROUND

Rounds numbers; those halfway between two integers are rounded up in magnitude: 2.5 is rounded to 3.0 while –2.5 is rounded to –3.0.

Syntax

ROUND   **E** to integer, put into **E**

Also see Arithmetic and Transformations.

RSSQ

Calculates the uncorrected sum of squares of the numbers in each row. If X_1, X_2, ..., X_8 are the numbers in row 1 of C1-C8, then RSSQ calculates $X_1^2 + X_2^2 = ... + X_8^2$. Missing observations are omitted.

Syntax

RSSQ   sum of the squared values per row of **E** . . . **E**, put in **C**

Also see Arithmetic and Transformations.

RSTDEV

Calculates the standard deviation of the numbers in each row, omitting missing observations.

Syntax

RSTDEV   standard deviation of all values per row of **E** . . . **E**, put in **C**

Also see Arithmetic and Transformations.

RSUM

Sums across the nonmissing values in each row, placing the result in the same row of column C.

Syntax

RSUM   sum of all values per row of **E** . . . **E**, put in **C**

Also see Arithmetic and Transformations.

RUNS

Performs a two-sided runs test.

Syntax

> **RUNS** test [above and below **K**] for data in **C**

How to Use

This nonparametric procedure does a two-sided runs test to check if the data are in random order. A run is a series of one or more observations in a row that are greater than K, or one or more observations in a row that are less than or equal to K. If K is not specified, the mean of the column is used.

Figure 6–49 shows a runs test performed on the inches of snowfall in a small midwestern town for the past 15 years.

FIGURE 6–49

```
STAT> RETRIEVE 'SNOWFALL'
STAT> RUNS 'SNOWFALL'

    SNOWFALL

    K =    40.0933

    THE OBSERVED NO. OF RUNS =    7
    THE EXPECTED NO. OF RUNS =    8.4667
        7 OBSERVATIONS ABOVE K    8 BELOW
 * N SMALL--FOLLOWING APPROX. MAY BE INVALID
            THE TEST IS SIGNIFICANT AT   0.4299
            CANNOT REJECT AT ALPHA = 0.05
```

STAT101 calculates the expected number of runs, and then it computes the probability that the observed number of runs will vary by this amount from the expected value.

SAMPLE

Randomly samples rows from one or more columns.

Syntax

> **SAMPLE** **K** rows from columns **C** ... **C**, put in **C** ... **C**

How to Use

With this command, you can sample K rows at random, without replacement, from a block of columns. In Fig. 6–50, rows 1, 5, and 4 were chosen from 'HT', 'WT', and 'CHANGE'.

FIGURE 6–50

```
STAT> RETRIEVE 'EXAMPLE'
STAT> SAMPLE 3 'HT' 'WT' 'CHANGE' C12-C14
STAT> PRINT 'HT' 'WT' 'CHANGE' C12-C14
```

ROW	HT	WT	CHANGE	C12	C13	C14
1	70	165	5	68	158	-2
2	62	110	-4	66	*	*
3	64	*	*	70	165	5
4	71	183	0			
5	66	*	*			
6	68	158	-2			
7	61	106	-7			

Your data will of course be different, as STAT101 will generate different values.

Also see RANDOM.

SAVE

Copies the current worksheet into a saved worksheet file.

Syntax

SAVE	['FILENAME']
PORTABLE	

How to Use

SAVE copies all the columns, data, column names, and constants from the current worksheet into a file with the default extension .MTW. For example, the command

⇨ ```
STAT> SAVE 'B:STATDATA'
Worksheet saved into file: B:STATDATA.MTW
```
```

creates a saved worksheet file STATDATA.MTW on the disk in drive B:. You can then use the command RETRIEVE to enter this file into the current worksheet of a STAT101 session on any DOS microcomputer. (For use on other computers, see the subcommand PORTABLE, in the subcommand section that follows.) The file cannot be sent to a printer or used by another software package. Use WRITE for that purpose.

File names may be any combination of letters and numbers up to eight characters long; blanks are not allowed. If you don't enter a file name, STAT101 saves your worksheet as STAT101.MTW. You may include an extension different from MTW in 'FILENAME'.

Subcommand

PORTABLE

This subcommand saves the worksheet in a file with the default extension .MTP, which can be used by STAT101 on any type of computer (e.g., a mainframe); just use RETRIEVE with the PORTABLE subcommand.

Also see RETRIEVE and WRITE.

SET

Enters data into one column. You can also enter data into the Data Editor; see Chapter 3.

Syntax

SET	data [from 'FILENAME'] into C
FORMAT	(format statement)
NOBS	= K

How to Use

With SET you can enter data into a column from the keyboard or from a data file. Use at least one space or a comma between data values.

FIGURE 6-51

```
STAT> SET C1
DATA>    6    1   *
DATA>    0   -2
DATA> END
```

In Fig. 6–51, the SET command puts the values 6, 1, *, 0, and –2 into column C1. The END command tells STAT101 you are finished entering data. As with READ (used for entering multiple columns), SET automatically erases the existing contents of the column listed, though it doesn't erase the column names. (To add numbers to existing columns, use INSERT.) A stored constant such as K1 may also be used as part of the data.

Entering Patterned Data Using SET

A useful function of the SET command is its ability to place patterned data into a column; it can save you a lot of typing. To set C3 with the numbers 1, 2, 3, 4, and 5, use a colon to abbreviate the range, as shown in Fig. 6–52:

FIGURE 6-52

```
STAT> SET C3
DATA>    1:5
DATA> END
STAT> PRINT C3

C3
    1    2    3    4    5
```

This list illustrates other conventions for setting columns with patterns:

4:1	expands to 4, 3, 2, 1
1:3/.5	expands to 1, 1.5, 2, 2.5, 3
3(1)	expands to 1, 1, 1
3(1:3)	expands to 1, 2, 3, 1, 2, 3, 1, 2, 3
(1:3)2	expands to 1, 1, 2, 2, 3, 3
3(1:3)2	expands to 1, 1, 2, 2, 3, 3, 1, 1, 2, 2, 3, 3, 1, 1, 2, 2, 3, 3

A slash indicates an increment. Put lists that you want repeated into parentheses. There must be no space between the repeat factor and the corresponding parenthesis. If the repeat factor is before the parentheses, STAT101 repeats the entire list; if it is after, STAT101 repeats each number in the range. You cannot nest parentheses. You can use a stored constant in place of any number on a data line after SET. Figure 6–53 is a more complicated example.

FIGURE 6–53

```
STAT> LET K1 = 2
STAT> SET C1
DATA>   5(1) 5(2) *
DATA>   2(1:3) (-2:0)2
DATA>   3(8:10)K1
DATA>   4:2/0.5
DATA> END
```

Figure 6–53 shows STAT101 storing 46 elements in C1. The first DATA> line enters 11 values: five 1's, five 2's and the missing value symbol (*). The second DATA> line produces 12 more rows: 2(1:3) expands as 1, 2, 3, 1, 2, 3, (two sets of the numbers 1 through 3), and (–2:0)2 expands as –2, –2, –1, –1, 0, 0 (each integer from –2 to 0 is entered twice). The third DATA> line stores another 18 values: the series 8, 8, 9, 9, 10, 10, entered three times. The last DATA> line enters five more numbers: 4, 3.5, 3, 2.5, 2.

Subcommands

FORMAT   (format statement)

Use this subcommand when you want to enter alpha data or need to give STAT101 special instructions. For example, to copy data into column C10 from a file TEST.DAT on the disk in drive B:, you could use the command:

⇨ STAT> SET 'B:TEST' C10;
   SUBC>   FORMAT (5F2.0)

Each line of the file contains five whole numbers, with two spaces per number. See online HELP SET FORMAT for more information on formatted input and output.

NOBS   = K

The NOBS subcommand tells STAT101 how many rows you will enter. The END command is not necessary if you use NOBS.

Also see READ, INSERT, and LET.

SIGNS

Converts negative numbers, zero, and positive numbers to −1, 0, and +1, respectively, and optionally stores these new values.

Syntax

SIGNS	of E, [put into E]

Also see Arithmetic and Transformations.

SIN

Calculates sines. The argument must be in radians. To convert from degrees to radians, multiply by 0.017453. STAT101 sets the result to *, the missing value symbol, and prints a "value out of bounds" message if the numbers are too large to be accurate. The exact magnitude of the out of bounds range depends on the computer.

Syntax

SIN	of E, put into E

Also see Arithmetic and Transformations.

SINTERVAL

Computes confidence intervals for the median.

Syntax

SINTERVAL	[confidence = K] on data in C . . . C

How to Use

This command calculates three confidence intervals around the median for each column listed: one with confidence just above K, one with confidence just below K, and a middle interval that approximates K. The middle interval in SINTERVAL, denoted by NLI, and the confidence interval displayed by BOXPLOT;NOTCH are calculated using a nonlinear interpolation procedure. If K is not specified, STAT101 uses 95%.

FIGURE 6-54

```
STAT> RETRIEVE 'STA261'
STAT> SINTERVAL 95 'FINAL'

SIGN CONFIDENCE INTERVAL FOR MEDIAN

                      ACHIEVED
               N    MEDIAN   CONFIDENCE    CONFIDENCE INTERVAL    POSITION
FINAL         45    73.00      0.9275       ( 69.00,    81.00)          17
                               0.9500       ( 68.54,    81.00)         NLI
                               0.9643       ( 68.00,    81.00)          16
```

In Fig. 6-54, the 17 under POSITION means that the first confidence interval goes from the 17th smallest value to the 17th largest one.

Also see STEST and BOXPLOT.

SORT

Sorts data in ascending order.

Syntax

SORT   data in C [carry along corresp. rows of C . . . C], put into C [corresp. rows into C . . . C]

How to Use

This command sorts data in a series of columns by rows, according to the values in the first column you list, then stores the rearranged data in the second series of columns, as shown in Fig. 6-55 on the following page.

FIGURE 6–55

```
STAT> RETRIEVE 'EXAMPLE'
STAT> SORT 'WT' 'HT' 'CHANGE' C12-C14
STAT> PRINT 'WT' 'HT' 'CHANGE' C12-C14
```

ROW	WT	HT	CHANGE	C12	C13	C14
1	165	70	5	106	61	-7
2	110	62	-4	110	62	-4
3	*	64	*	158	68	-2
4	183	71	0	165	70	5
5	*	66	*	183	71	0
6	158	68	-2	*	64	*
7	106	61	-7	*	66	*

For example, the SORT command in Fig. 6–55 sorts the rows in 'WT', 'HT', and 'CHANGE' by the values in 'WT' and stores them in columns C12 to C14. Note that * symbols are placed last.

Also see RANK.

SQRT

Calculates square roots. If you take the square root of a negative number, STAT101 sets the answer to *, missing value, and prints the "out of bounds" message.

Syntax

SQRT	of E, put into E

Also see Arithmetic and Transformations.

SSQ

Squares and sums the values in a column, omitting missing observations.

Syntax

SSQ	Sum of the squared values in **C** [put in **K**]

Also see **Arithmetic** and **Transformations**.

STACK

Stacks blocks of columns and/or constants on top of each other.

Syntax

STACK (E ... E) ... on top of (E ... E) put in (C ... C)

SUBSCRIPTS into C

How to Use

The first block is always placed on top of the second block, and so on, with results stored in the last block listed. In Fig. 6–56, the numbers in C16 show which block the values originally came from.

FIGURE 6–56

```
STAT> RETRIEVE 'EXAMPLE'
STAT> STACK (C8 C11) (C7 C9) (14 8) (C14 C15);
SUBC>    SUBSCRIPTS C16.
STAT> PRINT C8 C11 C7 C9 C14-C16
```

ROW	C8	C11	C7	C9	C14	C15	C16
1	1	5	0	5	1	5	1
2	2	6	1	6	2	6	1
3	3	7			3	7	1
4	4	8			4	8	1
5					0	5	2
6					1	6	2
7					14	8	3

Columns within each block must be of the same length. You may omit the parentheses around each block if your blocks are only one column (or constant) wide.

Subcommand

SUBSCRIPTS into C

This subcommand stores an extra column of numbers to show which block each row came from. Rows from the first block are assigned the value of 1, the second block a value of 2, and so on, as shown in Fig. 6–56.

Also see UNSTACK.

STDEV

Calculates the standard deviation of C.

Syntax

STDEV	standard deviation of **C** [put in **K**]

Also see Arithmetic and Transformations.

STEM-AND-LEAF

Constructs a histogram with digits from the actual data values.

Syntax

STEM-AND-LEAF	of the data in **C . . . C**
TRIM	outliers
INCREMENT	= **K**
BY	**C**

How to Use

The STEM-AND-LEAF display has three columns. The column on the left is a cumulative count of values from the top of the figure down and the bottom of the figure up to the middle. The number in parentheses is the count of values in the row containing the median. Each number in the second column is called a stem, while the numbers to its right are the leaves, each of which is a single digit to place after the stem digits, representing one data value. The leaf unit tells you where to put the decimal place in each number, as shown in Fig. 6–57.

FIGURE 6-57

```
STAT> RETRIEVE 'STA261'
STAT> STEM 'FINAL'

Stem-and-leaf of FINAL    N = 45
Leaf Unit = 1.0

    1     0  0
    1     1
    1     2
    3     3  48
    4     4  7
    7     5  359
   18     6  12356666899
   (7)    7  0011378
   20     8  11111225555899
    6     9  001258
```

In the stem-and-leaf display in Fig. 6-57, each leaf digit is in the ones position. Thus the lowest grade was a 0, followed by 34, 38, 47, and so on. The median, 73, is the fifth value in the row, the count of which is enclosed by parentheses. The highest grade was a 98.

Parentheses around the median row are omitted if the median falls between two lines of the display.

Subcommands

TRIM   outliers

This subcommand omits outliers (see BOXPLOT for a definition) from the stem-and-leaf display, and shows them on lines labeled LO and HI. TRIM and BY cannot be used together.

INCREMENT   = K

This controls vertical scaling. The value of K will be the difference between the smallest possible values on adjacent lines.

BY C

BY produces a stem-and-leaf display for each distinct value in C, which may contain whole numbers from −9999 to +9999 or the missing value symbol (*). This is useful if the column C contains codes for subgroup membership (e.g., 1's and 2's for males and females). BY cannot be used with TRIM.

Also see DOTPLOT, HISTOGRAM, and BOXPLOT.

STEPWISE

Performs a stepwise regression to find the best subset of predictors with respect to the F-statistic.

Syntax

STEPWISE	regression of y in C on the predictors C . . . C
FENTER	= K
FREMOVE	= K
FORCE	C . . . C
ENTER	C . . . C
REMOVE	C . . . C
BEST	K alternative predictors
STEPS	= K

How to Use

This command does a linear regression step by step in a search of a good subset of the possible predictors. You can vary the basic procedure to allow for forward selection or for backwards elimination. Unless you are running STEPWISE from a command file, you can also intervene at any pause to modify the procedure with new subcommands. Models are fit with a constant term unless the NOCONSTANT command is in effect.

Default Procedure

The basic method alternates between stages of adding and removing variables until no more predictors can enter or leave the model. In step one, STAT101 calculates an F-statistic for each predictor already in the model. If that value is less than FREMOVE (4, unless otherwise specified) for any predictor, the predictor with the lowest F-statistic is removed, and output from the resulting model is printed. In step two, STAT101 calculates an F-statistic for each predictor not in the current model. If that value is greater than FENTER (4, unless otherwise specified) for any predictor, the predictor with the highest F-statistic is entered, and output from the resulting model is printed.

Forward Selection

This method adds predictors to the model as in the default procedure, but never removes any. To do forward selection, set FREMOVE to zero.

Backwards Elimination

This method starts with a model containing all possible predictors, then removes them one at a time without ever reentering any. To do backwards elimination, set FENTER = 100000 and list all predictors on the ENTER subcommand.

User Intervention

After each screenful of STEPWISE output, STAT101 displays the prompt MORE?. You may respond YES to continue the procedure, NO to terminate it, or you may alter the model with one or more of the subcommands described below. (The STEPS subcommand lets you control the amount of output in one screenful.)

Output

STAT101 prints the constant term (if applicable), the coefficient and t-ratio for each predictor, and the s and R-squared values for each step.

FIGURE 6-58

```
STAT> RETRIEVE 'STA261'
STAT> STEPWISE 'FINAL' C1-C14

STEPWISE REGRESSION OF  FINAL   ON 14 PREDICTORS, WITH N =    45
    STEP         1        2        3        4
CONSTANT     6.747   -7.132  -15.094  -15.487

T3           0.886    0.632    0.589    0.642
T-RATIO       9.41     5.24     5.19     6.00

T2                    0.43     0.38     0.39
T-RATIO               3.01     2.80     3.12


Q2                             1.23     1.36
T-RATIO                        2.76     3.28

Q6                                     -0.67
T-RATIO                                -2.75

S            10.5     9.65     8.97     8.33
R-SQ        67.33    73.13    77.34    80.93
  MORE? (YES, NO, SUBCOMMAND, OR HELP)
SUBC>    NO
```

In Fig. 6-58, the response to the MORE? prompt was NO. Alternatively, the responses shown in Fig. 6-59

FIGURE 6-59

```
MORE? (YES, NO, SUBCOMMAND, OR HELP)
SUBC>    FENTER = 1;
SUBC>    FREMOVE = 1.
```

would cause STAT101 to compute and display five more steps, while lowering the required F-values.

Subcommands

FENTER   = K

With this subcommand, you can change the minimum F-value required to enter a predictor variable into a model from the default value of 4. The value for FENTER must be greater than or equal to FREMOVE.

FREMOVE	= K

This allows you to set the F-value below which a predictor variable will be removed from a model. The default value is 4.

FORCE	C . . . C

Here you list predictor variables that you want included in all subsequent models, until you change their status with a REMOVE or an ENTER subcommand.

ENTER	C . . . C

These variables are to be entered in the current model. Any of them may be removed in the next step if its F-value falls below FREMOVE.

REMOVE	C . . . C

These variables are to be removed from the current model. Any of them may reenter the model at the next step if its F-value rises above FENTER.

BEST	K alternative variables

For each step in which a variable is added to the model, the next K best alternatives (in terms of the F-statistic) are printed, along with their associated t-statistics. Each t-value shown is obtained by replacing the variable actually chosen by the alternative variable. The default value of K is 0.

STEPS	= K

You can set the number of steps of output printed before the MORE? prompt allows you to intervene again. K can be any number from 1 up to the maximum allowed by the width of your screen (or by the OW command).

Also see REGRESS, CONSTANT, NOCONSTANT, and OW.

STEST

Computes a sign test for one or more columns.

Syntax

STEST   [median = K] on data in C . . . C

ALTERNATIVE   = K

How to Use

Does a nonparametric, two-sided sign test of whether or not the median for each column listed is equal to the value K (unless you use the ALTERNATIVE subcommand).

FIGURE 6–60

```
STAT> RETRIEVE 'STA261'
STAT> STEST 13.5 'Q1' 'Q2'

SIGN TEST OF MEDIAN = 13.50 VERSUS N.E. 13.50

         N   BELOW  EQUAL  ABOVE  P-VALUE  MEDIAN
Q1      45    12      0     33    0.0025   14.00
Q2      45    23      0     22    1.0000   13.00
```

The example in Fig. 6–60 tests whether the median score on quizzes 1 and 2 is equal to 13.5. STAT101 prints the median, the number of observations below, equal to, and above the median, and the p-value or probability based on the binomial distribution.

The default value of K is 0. Any values equal to K are omitted from the sign test.

Subcommand

ALTERNATIVE   = K

Does a one-sided test of the alternative hypothesis H1:(median < K) for ALTERNATIVE = –1, or H1:(median > K) for ALTERNATIVE = 1. The null hypothesis in both cases is H0:(median = K). ALTERNATIVE = 0 performs the same two-sided test as STEST with no subcommand.

Also see SINTERVAL.

STOP

Exits from the STAT101 program.

Syntax

STOP

How to Use

STOP ends your STAT101 session and erases the current worksheet. Be sure to SAVE any data you may want to reuse before typing STOP.

Also see SAVE.

STORE

Stores subsequent commands in a command file.

Syntax

STORE   [in **'FILENAME'**] the following STAT101 commands

How to Use

A command file, or macro, contains STAT101 commands that can be processed with the EXECUTE command. STORE tells STAT101 to enter all subsequent commands into the specified file (its default name is STAT101.MTB) until it encounters the command END. Macros are given the default file extension .MTB (other extensions are permissible); they are standard ASCII text files that may be edited or sent to a printer.

Figure 6–61 shows how to store a confidence interval simulation macro:

FIGURE 6-61

```
STAT> STORE 'CONFINTS'
STOR> RANDOM 30 C1-C25:
STOR> NORMAL 10 4.
STOR> ZINTERVAL 90 4 C1-C25.
STOR> END
```

To execute this macro, use the EXECUTE command with the file name CONFINTS.

Also see EXECUTE, END, ECHO, NOECHO, and NOTE.

SUBTRACT

Does row-by-row subtraction on columns and/or constants. If you subtract columns of unequal length, STAT101 uses the shortest column as the length of the answer column.

Syntax

SUBTRACT   E from E, put into E

Also see Arithmetic and Transformations.

SUM

Adds values in a column.

Syntax

SUM   add all values in **C** [put in **K**]

Also see Arithmetic and Transformations.

SYSTEM

Temporarily exits from your STAT101 session to DOS.

Syntax

SYSTEM

How to Use

SYSTEM allows you to leave STAT101 temporarily in order to execute DOS commands (e.g., MKDIR, FORMAT, PRINT) and then return to your STAT101 session by typing EXIT.

It is safest to SAVE your worksheet and close an open outfile (NOOUT-FILE) before using the SYSTEM command. Also, never change your DOS path if you wish to return to a suspended STAT101 session.

Also see DIR, TYPE, and CD.

TABLE

Produces one-way or multi-way tables. The TABLE command prints one-way, two-way, and multi-way tables of categorical data. The cells may contain counts, percents, and statistics from a chisquare test. They may also contain summary statistics such as means, standard deviations, and maximums for associated variables.

Syntax

TABLE	the classification variables in C . . . C
MEANS	for C . . . C
MEDIANS	for C . . . C
SUMS	for C . . . C
MINIMUMS	for C . . . C
MAXIMUMS	for C . . . C
STDEV	for C . . . C
STATS	for C . . . C
DATA	for C . . . C

(continued)

N	for C ... C
NMISS	for C ... C
PROPORTION	of cases = K [thru K] in C ... C
COUNTS	
ROWPERCENTS	
COLPERCENTS	
TOTPERCENTS	
CHISQUARE	analysis [output code = K]
MISSING	level for the classification variable C ... C
NOALL	in margins
ALL	for the classification variables C ... C
FREQUENCIES	are in C

How to Use

To display STAT101 data in table form, enter the TABLE command, followed by up to ten classification variables (e.g., sex, year, group number). The values in the first column listed determine the row headings of the table, and the values in the second column determine the column headings. A separate table is produced for every possible combination of values from the remaining columns listed. They may contain whole numbers between −9999 and +9999 or the missing value symbol (∗). Variables listed in subcommands (except for MISSING and ALL) may be of any type (discrete or continuous). If no subcommands are used, STAT101 prints the number of observations in each cell of the table.

The example in Fig. 6–62 tabulates the number of observations and the average GPA for students who do and do not watch soap operas, and for those who do and do not have a boyfriend.

FIGURE 6-62

```
STAT> RETRIEVE 'MIAMI'
STAT> TABLE 'SOAPS' 'BOYFRND';
SUBC>    COUNTS;
SUBC>    MEAN 'GPA'.

     ROWS: SOAPS      COLUMNS: BOYFRND

                  1         2       ALL

        1        11        12        23
              3.0909    3.0417    3.0652

        2        37        23        60
              3.3514    3.3261    3.3417

      ALL        48        35        83
              3.2917    3.2286    3.2651

         CELL CONTENTS --
                          COUNT
                       GPA:MEAN
```

STAT101 prints subtotals (marginal values) across rows in a row labeled ALL and down columns in a column labeled ALL.

To list actual GPA values instead of the count and average for each cell, you could change the subcommands given in Fig. 6-63 as follows:

FIGURE 6-63

```
STAT> TABLE 'SOAPS' 'BOYFRND';
SUBC>    DATA 'GPA'.
```

Subcommands

MEANS   for C ... C

Prints the mean for each cell and the overall marginal means for each row and column.

MEDIANS   for C ... C

Prints the median for each cell; marginal medians are not printed.

| SUMS | for C ... C |

Prints the sum of the data for each cell.

| MINIMUMS | for C ... C |

Prints the minimum value for each cell.

| MAXIMUMS | for C ... C |

Prints the maximum value for each cell.

| STDEV | for C ... C |

Prints the standard deviation for each cell.

| STATS | for C ... C |

Prints the number of nonmissing values, the mean, and the standard deviation for each cell.

| DATA | for C ... C |

Prints all the data values for each cell.

| N | for C ... C |

Prints the number of nonmissing data values for each cell.

| NMISS | for C ... C |

Prints the number of missing values for each cell.

PROPORTION of cases = K [thru K] in C ... C

Prints the proportion of observations equal to K or in the range from K to K for each cell. There should not be any missing value symbols (*) in columns listed with this subcommand.

COUNTS

Prints the total number of values (missing and nonmissing) for each cell.

ROWPERCENTS

Calculates what percentage each cell represents of the total observations in the row.

COLPERCENTS

Calculates what percentage each cell represents of the total observations in the column.

In Fig. 6–64, STAT101 generates a table that shows the relationship between those who smoke (C4 SMOKES) and their activity level (C8 ACTIVITY). The COLPERCENTS subcommand gives the percentage of the entire column that the count represents:

FIGURE 6–64

```
STAT> RETRIEVE 'PULSE'
STAT> TABLE 'SMOKES' 'ACTIVITY';
SUBC>    COLPERCENTS.

  ROWS: SMOKES      COLUMNS: ACTIVITY

             0        1        2        3       ALL
     1   100.00    33.33    31.15    23.81    30.43
     2     --      66.67    68.85    76.19    69.57
   ALL   100.00   100.00   100.00   100.00   100.00

     CELL CONTENTS --
                 % OF COL
```

This table summarizes the number and percentage of smokers at each activity level. The 1's represent those who smoke regularly while the 2's are those who do not. A third of the inactive students smoke while only a fourth of the very active smoke. Further analysis would be necessary to test whether there is evidence that this is a significant difference.

TOTPERCENTS

Calculates what percentage each cell represents of all the observations in the table.

CHISQUARE   analysis [output code = K]

Prints the chisquare statistic, used to test the independence of the rows and columns in a two-way table, together with its associated degrees of freedom. The following data are printed in each cell, depending on the value of the output code K (its default value is 1):

K = 1   Observed counts for each cell.

K = 2   Observed and expected counts for each cell.

K = 3   Observed and expected counts and the standardized residuals for each cell, which are computed as (observed count − expected count) divided by the square root of the expected count.

FIGURE 6–65

```
STAT> RETRIEVE 'PULSE'
STAT> TABLE 'SEX' 'ACTIVITY';
SUBC>    CHISQUARE 3.
   ROWS: SEX      COLUMNS: ACTIVITY

               0         1         2         3        ALL
    1          1         5        35        16         57
            0.62      5.58     37.79     13.01      57.00
            0.48     -0.24     -0.45      0.83        --

    2          0         4        26         5         35
            0.38      3.42     23.21      7.99      35.00
           -0.62      0.31      0.58     -1.06        --

   ALL         1         9        61        21         92
            1.00      9.00     61.00     21.00      92.00
             --        --        --        --         --

 CHISQUARE =      3.118   WITH D.F. =     3
    CELL CONTENTS --
                   COUNT
                   EXP FREQ
                   STD RES
```

The CHISQUARE subcommand output also gives you a summary chisquare statistic (3.118 in Fig. 6–65) that you can compare with percentiles from the chisquare distribution, using STAT101's INVCDF distribution command.

MISSING   level for the classification variables **C ... C**

Adds an additional row or column labeled MISSING to the display, containing counts of the missing observations in each cell for any variables (from the TABLE command line) that are listed after this subcommand and have missing values.

NOALL   in margins

This subcommand cancels the printing of marginal values in rows and columns labeled ALL.

ALL   for the classification variables **C ... C**

Prints marginal values only for the classification variables from the TABLE command line that are listed with this subcommand.

FREQUENCIES   are in **C**

The TABLE command normally assumes that each row in the input columns refers to one case. Sometimes you may find it necessary to combine the raw-data values into partial or full-frequency counts. This subcommand tells STAT101 where to find the frequencies (or weights) for each row.

Also see CHISQUARE and TALLY.

TALLY

Tallies information for each value in the columns listed.

Syntax

TALLY	the data in C . . . C
COUNTS	
PERCENTS	
CUMCOUNTS	cumulative counts
CUMPERCENTS	cumulative percentages
ALL	four statistics above

How to Use

The columns listed with TALLY must represent classification variables (e.g., sex, group number, year) and contain only whole numbers between −9999 and +9999 or missing value symbols (*). If no subcommands are used, TALLY prints only the counts for each value in the columns listed.

In Fig. 6–66, the values 1, 2, and 3 represent short, medium-length, and long hair respectively. TALLY shows that 34 individuals (about 41% of the total) have short hair, 39 (about 47%) have medium-length hair, and 10 (about 12%) have long hair. Cumulative counts and percentages are also displayed.

FIGURE 6-66

```
STAT> RETRIEVE 'MIAMI'
STAT> TALLY  'HAIR LNG';
SUBC>   ALL.

HAIR LNG   COUNT  CUMCNT  PERCENT   CUMPCT
       1      34      34    40.96    40.96
       2      39      73    46.99    87.95
       3      10      83    12.05   100.00
      N=      83
```

Subcommands

COUNTS

Prints the number of times each distinct value occurs in the columns listed. This information is printed by default if you don't list any subcommands.

PERCENTS

Prints the relative frequency of each nonmissing value in the columns listed.

CUMCOUNTS

Prints cumulative counts of the nonmissing values in each column listed.

CUMPERCENTS

Prints cumulative relative frequencies of the nonmissing values in each column listed.

ALL

Prints counts, cumulative counts, percentages, and cumulative percentages for the distinct value in each column listed.

Also see TABLE.

TAN

Calculates tangents. The argument must be in radians. To convert from degrees to radians, multiply by 0.017453. STAT101 sets the result to *, the missing value symbol, and prints a "value out of bounds" message if the numbers are too large to be accurate. The exact magnitude of the out of bounds range depends on the computer.

Syntax

TAN   of E, put into E

Also see Arithmetic and Transformations.

TINTERVAL

Calculates a confidence interval for the mean of each column

Syntax

TINTERVAL   [K% confidence] for data in **C** . . . **C**

How to Use

With this command you can calculate a confidence interval for the mean of one or more variables (assumed to be normal) without knowing their standard deviations. The procedure uses values from the Student's t-distribution, and it assumes K = 95% as the default confidence level. Note that K may be expressed either as a percentage or as a decimal, such as .95. Figure 6–67 is an example of a TINTERVAL test on one column of data.

FIGURE 6–67

```
STAT> RETRIEVE 'EXAMPLE'
STAT> TINTERVAL 90 'HT'

            N      MEAN    STDEV    SE MEAN    90.0 PERCENT C.I.
HT          7     66.00     3.87       1.46    (   63.15,    68.85)
```

Also see TTEST, ZINTERVAL, ZTEST, SINTERVAL, and WINTERVAL.

TPLOT

Constructs a pseudo-three-dimensional plot.

Syntax

TPLOT   y in **C** vs x in **C** vs z in **C**

YINCREMENT   = K

YSTART   at K [end at K]

XINCREMENT	= K
XSTART	at K [end at K]

How to Use

This command plots the data from three columns. The first column is represented on the y-axis of the plot, the second column is represented on the x-axis of the plot, and the third, or z, column is represented with a set of special plotting symbols.

The plotting symbol used for z is "0" if z is more than one standard deviation below its mean, "." if z is within one standard deviation below the mean, "/" if z is within one standard deviation above the mean, and "X" if z is more than one standard deviation above the mean.

FIGURE 6-68

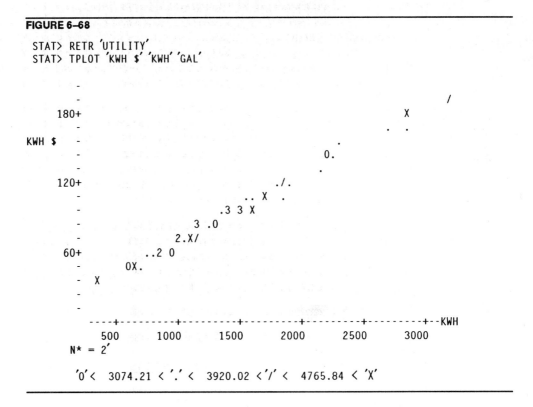

If several points fall on the same spot, as in Fig. 6–68, STAT101 prints a count instead. Use the commands HEIGHT and WIDTH to control the size of the TPLOT, if desired. You may override STAT101's default scaling with the subcommands, which are described further under the command PLOT.

Also see PLOT, LPLOT, MPLOT, HEIGHT, and WIDTH.

TSPLOT

Plots time-series data.

Syntax

TSPLOT	[with period K] time series data in C
INCREMENT	= K
START	at K [end at K]
ORIGIN	= K
TSTART	at K [end at K]

How to Use

This command plots the data on the vertical axis against time on the horizontal axis, with the numbers 1 through K (period length) as plotting symbols. For example, if you specify a period length of 5, each group of five observations will be plotted with the numbers 1, 2, 3, 4, and 5. For a period length greater than 9, data are plotted with a 0 for 10, A for 11, ... , Z for 36. The period sets the distance between the tick marks (+) on the horizontal axis. It may be any positive integer up to 36; the default is 10. See Fig. 6–69 as follows:

FIGURE 6-69

```
STAT> RETRIEVE 'UTILITY'
STAT> TSPLOT 12 'KWH'
```

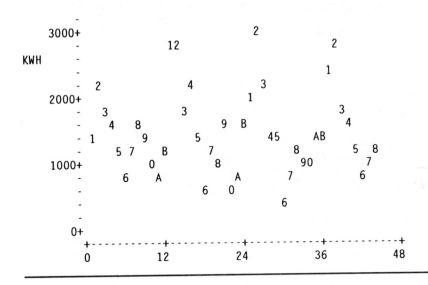

Use the commands HEIGHT and OW (not WIDTH) to control the size of the TSPLOT if you wish. You may override STAT101's default scaling of TSPLOT with the subcommands described below.

Subcommands

INCREMENT = K

This subcommand controls vertical scaling: K is the distance between the tick marks (+) on the y-axis.

START at K [end at K]

This subcommand specifies the first and, optionally, the last value to print on the vertical scale of the TSPLOT. Data outside this range are excluded.

ORIGIN = K

This subcommand tells STAT101 the scale of numbers to use when labeling the horizontal axis. The first point will be plotted directly above K. For example, if your series consisted of annual figures starting in the year 1950, you might enter ORIGIN = 1950.

TSTART = K [end at K]

This subcommand allows you to graph a subset of your time series by specifying the first and, optionally, the last observation to plot. For example, TSTART = 4 begins the TSPLOT with the fourth observation. If you specify a scale with the ORIGIN subcommand, any values of K with TSTART must be on the same scale.

Also see PLOT, LPLOT, MPLOT, TPLOT, OW, and HEIGHT.

TTEST

Performs a t-test for the mean of each column listed.

Syntax

TTEST [μ= K] on C ... C

ALTERNATIVE = K

How to Use

This command does a two-sided t-test of the null hypothesis H_0:(μ = K) versus H_1:($\mu \neq$ K) on each column of data, where the standard deviation is not known. The default value of K is 0. TTEST also computes the probability (p) of getting a value as extreme as, or more extreme than, the computed t-value from a t-distribution with (n − 1) degrees of freedom. You can use the ALTERNATIVE subcommand to do a one-sided test.

Subcommand

ALTERNATIVE = K

Does a one-sided test of the alternative hypothesis $H_1:(\mu < K)$ for ALTERNATIVE = –1, or $H_1:(\mu > K)$ for ALTERNATIVE = 1. The null hypothesis is $H_0:(\mu = K)$ for both cases. (ALTERNATIVE = 0 does the same test as TTEST without a subcommand.) In Fig. 6–70, STAT101 tests whether the mean final exam grade is different from 75.

FIGURE 6–70

```
STAT> RETRIEVE 'STA261'
STAT> TTEST 75 'FINAL';
SUBC>    ALTERNATIVE 1.

TEST OF MU = 75.000 VS MU G.T. 75.000

             N      MEAN    STDEV   SE MEAN       T    P VALUE
FINAL       45    72.178   18.189     2.711   -1.04      0.85
```

Paired Data

To do a t-test on paired data, use LET or SUBTRACT to store the difference between the two paired columns in a third column. Then use TTEST on that third column.

Also see TINTERVAL, ZTEST, ZINTERVAL, and SINTERVAL.

TWOSAMPLE

Compares the means of two populations with a t-test, and computes a confidence interval for the difference.

Syntax

TWOSAMPLE   [K% confidence] for data in C and C

ALTERNATIVE   = K

POOLED

How to Use

This command does a t-test of the null hypothesis $H_0:(\mu_1 = \mu_2)$ versus $H_1: (\mu_1 \neq \mu_2)$ on two columns of data of any length, and finds a confidence interval for $(\mu_1 - \mu_2)$. It computes the probability (P) of getting a value as as, or more extreme than, the computed t-value from a t-distribution with (n – 1) degrees of freedom. TWOSAMPLE is illustrated in Fig. 6–71.

FIGURE 6-71

```
STAT> RETRIEVE 'CARSPEED'
STAT> TWOSAMPLE 'METHOD 1' 'METHOD 4'

TWOSAMPLE T FOR METHOD 1 VS METHOD 4
            N      MEAN     STDEV    SE MEAN
METHOD 1   12     64.42      7.10      2.1
METHOD 4   11     55.36      4.46      1.3

95 PCT CI FOR MU METHOD 1 - MU METHOD 4: (3.9, 14.2)

TTEST MU METHOD 1 = MU METHOD 4 (VS NE): T= 3.69   P=0.0017   DF=  18
```

TWOSAMPLE differs from the usual two-sample test in that it does not calculate a t-statistic based on an assumption that the two populations have equal variances. To do the usual pooled t-test, which assumes the populations have equal variances, use the POOLED subcommand.

You may specify a confidence K either as a percentage or as a decimal such as 0.90; the default is 95%. You can use the ALTERNATIVE subcommand to do a one-sided test.

Subcommands

ALTERNATIVE = K

Does a one-sided test of the alternative hypothesis $H_1:(\mu_1 < \mu_2)$ for ALTERNATIVE = –1, or $H_1: (\mu_1 > \mu_2)$ for ALTERNATIVE = 1. The null hypothesis is $H_0:(\mu_1 = \mu_2)$ in both cases. ALTERNATIVE = 0 does the same test as TWOSAMPLE without the ALTERNATIVE subcommand.

POOLED

Tells STAT101 to assume the two populations have a common variance.

Also see TTEST, TWOT, AOVONEWAY, and MANN-WHITNEY.

TWOT

Compares the means of two populations with a t-test, and computes a confidence interval for their difference.

Syntax

TWOT	[K% confidence] for data in C, subscripts in C
ALTERNATIVE	= K
POOLED	

How to Use

TWOT does exactly the same tests and confidence-interval computation as the TWOSAMPLE command. The only difference is in where STAT101 looks for the data. With TWOSAMPLE, sample data from two populations are in two separate columns. With TWOT, all sample data are in the first column listed. The corresponding row in the second column tells STAT101 which of the two populations each value belongs to. Subscripts may be any integers from −9999 to +9999 or the missing value symbol (*). See Fig. 6–72 for an example.

FIGURE 6–72

```
STAT> RETRIEVE 'MIAMI'
STAT> TWOT 'CREDITS' 'JOB'

TWOSAMPLE T FOR CREDITS
JOB   N      MEAN     STDEV    SE MEAN
 1    41     16.78     1.84     0.29
 2    42     17.05     1.87     0.29

95 PCT CI FOR MU 1 - MU 2: (-1.08, 0.54)

TTEST MU 1 = MU 2 (VS NE): T= -0.66  P=0.51  DF= 80
```

See TWOSAMPLE for further explanation of this test and of the subcommands. (You can use STACK or UNSTACK to combine or separate data sets.)

Also see TTEST, TWOSAMPLE, and ONEWAY.

TWOWAY

Performs a two-way analysis of variance for balanced data.

Syntax

TWOWAY	for the data in **C**, subscripts in **C** and **C** [store residual values in **C** [fits in **C**]]
ADDITIVE	
MEANS	for the subscripts in **C** [**C**]

How to Use

In a two-way analysis of variance, data are classified by two sets of factors representing treatments or blocks, each having more than one level. The number of levels in factor one times the number of levels in factor two equals the number of cells, which represents the number of different factor-level combinations. Each of these cells must contain an equal number of observations (i.e., the data must be balanced).

The first column listed with TWOWAY should contain the sample data. The corresponding row in the second (factor one) and third (factor two) columns listed tells STAT101 which cell each data value belongs to. Specify additional columns with TWOWAY if you want to store residuals and fitted values. See Fig. 6–73.

FIGURE 6–73

```
STAT> READ C1-C3
DATA>    47    2    1
DATA>    43    2    1
DATA>    46    2    2
DATA>    40    2    2
DATA>    62    3    1
DATA>    68    3    1
DATA>    67    3    2
DATA>    71    3    2
DATA>    41    4    1
DATA>    39    4    1
DATA>    42    4    2
DATA>    46    4    2
DATA> END
       12 ROWS READ
STAT> TWOWAY C1 C2 C3;
SUBC>    MEAN C2 C3.
```

```
ANALYSIS OF VARIANCE  C1
SOURCE          DF      SS        MS
C2               2    1544.0     772.0
C3               1      12.0      12.0
INTERACTION      2      24.0      12.0
ERROR            6      62.0      10.3
TOTAL           11    1642.0
```

```
                           Individual 95% CI
     C2      Mean     ---+---------+---------+---------+--------
      2      44.0        (----*----)
      3      67.0                                           (----*----)
      4      42.0     (---*----)
                      ---+---------+---------+---------+--------
                       40.0      48.0      56.0      64.0

                           Individual 95% CI
     C3      Mean     ---+---------+---------+---------+--------
      1      50.0     (------------*------------)
      2      52.0                 (------------*------------)
                      ---+---------+---------+---------+--------
                       47.5      50.0      52.5      55.0
```

After you compute the appropriate F-statistics (which depend on whether the factors are fixed or random), you can use STAT101 to find the corresponding critical F-value with the INVCDF command. First choose a level of significance α. Then type INVCDF with $1 - \alpha$, entering the appropriate degrees of freedom for the numerator and denominator of F. For example, to find the critical F for group C2 with $\alpha = 0.05$, use the command:

⇨ STAT> INVCDF 0.95;
  SUBC>   F 2 6.
      0.9500   5.1433

Factor columns may contain any integers from −9999 to +9999 or missing values (∗). An interaction term will automatically be included in the model if there is more than one observation per cell. To produce cell means and standard deviations, use the TABLE command.

For a quick test to see whether or not your data are balanced, use the TABLE command with your two factor columns. The number of observations per cell will be displayed.

Subcommands

ADDITIVE

This subcommand tells STAT101 to fit a model without an interaction term. In this case, the fitted value for cell (i,j) is (mean of observations in row i) + (mean of observations in row j) − (mean of all observations).

MEANS   for the factors C [C]

This subcommand calculates marginal means and 95% confidence intervals for each level of either or both factor columns. Each confidence interval is calculated from the Student's t-distribution, using $\overline{\text{MS ERROR}}$ degrees of freedom and a pooled standard deviation equal to $\sqrt{\text{MS ERROR}}$.

Also see ONEWAY, TABLE, INVCDF, and CDF.

TYPE

Lists the contents of an ASCII file.

Syntax

TYPE   file name

How to Use

You can use the TYPE command in STAT101 just as in DOS to display the contents of a file on your screen. The file name should contain path information (disk drive and directory name) if necessary, and should not be enclosed in single quotes. Any standard format (ASCII) text file can be displayed. For example, Fig. 6–74 on the next page shows creating an outfile of a class assignment in which you retrieve the MIAMI data set and print basic statistics about C1.

FIGURE 6-74

```
STAT> OUTFILE 'ASSIGN'
STAT> #NAME:      JOAN CAREY
STAT> #CLASS:     STAT101
STAT> #PROFESSOR: DR. GILLETT
STAT> #TIME:      MWF 9:50
STAT> RETRIEVE 'MIAMI'
   WORKSHEET SAVED  3/ 3/1993

Worksheet retrieved from file: MIAMI.MTW
STAT> DESC C1

              N       MEAN    MEDIAN    TRMEAN    STDEV    SEMEAN
HAIR         83      1.590     1.000     1.493    0.976     0.107

            MIN        MAX        Q1        Q3
HAIR      1.000      4.000     1.000     3.000

STAT> NOOUTFILE
```

Use this command to type the outfile ASSIGN.LIS on the screen:

⇨   STAT> TYPE 'ASSIGN.LIS'

The output that appears on the screen reproduces the commands and output produced while you were creating the outfile. If you printed ASSIGN.LIS on a printer, you could hand it in to your professor.

Also see CD, DIR, and OH.

UNSTACK

Separates one block of columns into several shorter blocks of columns or constants.

Syntax

UNSTACK	(C ... C) into (E ... E) ... (E ... E)
SUBSCRIPTS	are in C

How to Use

UNSTACK separates the first block of columns by rows into several blocks of new columns or constants, according to the values in the SUBSCRIPTS column. Rows with the smallest subscript are stored in the first block, rows with the next smallest subscript are stored in the second block, and so on. All blocks must be enclosed in parentheses unless they are only one column wide. Using the EXAMPLE.MTW saved worksheet, for example, use UNSTACK to store male ('SEX' = 1) heights and weights in columns C11 and C12, and female ('SEX' = 2) heights and weights in columns C13 and C14, as shown in Fig. 6–75.

FIGURE 6–75

```
STAT> RETRIEVE 'EXAMPLE'
STAT> UNSTACK ('HT' 'WT') (C11 C12) (C13 C14);
SUBC>     SUBSCRIPTS 'SEX'.
STAT> PRINT 'SEX' 'HT' 'WT' C11-C14
```

ROW	SEX	HT	WT	C11	C12	C13	C14
1	1	70	165	70	165	62	110
2	2	62	110	64	*	66	*
3	1	64	*	71	183	61	106
4	1	71	183	68	158		
5	2	66	*				
6	1	68	158				
7	2	61	106				

Figure 6–76 shows creating a column containing every second row of the column 'HT', which has seven rows. With SET you can make a column of distinguishing subscripts for UNSTACK to use.

FIGURE 6–76

```
STAT> SET C12
DATA>    1  2  1  2  1  2  1
DATA> END
STAT> UNSTACK 'HT' C13 C14;
SUBC>     SUBSCRIPTS C12.
STAT> PRINT 'HT' C12-C14
```

ROW	HT	C12	C13	C14
1	70	1	70	62
2	62	2	64	71
3	64	1	66	68
4	71	2	61	
5	66	1		
6	68	2		
7	61	1		

This example puts alternating 1's and 2's into column C6, and then uses these subscripts to store every second row of 'HT' in column C14.

If the SUBSCRIPTS command is not used, STAT101 assumes that each row from the first column should be put in a separate block, as shown in Fig. 6–77. (First use the RESTART command to clear the worksheet, and then retrieve a clean copy of the EXAMPLE data set.)

FIGURE 6–77

```
STAT> RESTART
STAT> RETRIEVE 'EXAMPLE'
STAT> PRINT C11

C11
    5    6    7    8

STAT> UNSTACK C11 K1-K4
STAT> PRINT C11 K1-K4
K1       5.00000
K2       6.00000
K3       7.00000
K4       8.00000

C11
    5    6    7    8
```

In Fig. 6–77, UNSTACK stores the values in C11 in constants K1 through K4.

Subcommand

SUBSCRIPTS are in C

The values in column C tell STAT101 which block of columns of constants to store unstacked data in. The subscripts must be integers from −9999 to +9999 or missing values (∗).

Also see STACK and SET.

WALSH

Calculates the averages of all pairs of values from one column.

Syntax

WALSH averages for C, put into C [put indices into C and C]

How to Use

This command, useful for nonparametric tests, calculates the average for each possible pair of values from the column, including the average of each value with itself. For n data values, there will be $n(n+1)/2$ Walsh averages stored. If you list two extra output columns, the indices i and j for each average $(X_i + X_j)/2$ will also be stored, as shown in Fig. 6–78.

FIGURE 6–78

```
STAT> SET C1
DATA>    77 88 85
DATA> END
STAT> WALSH C1 C2 C3 C4
STAT> PRINT C1-C4
 ROW     C1      C2     C3    C4
  1      77     77.0    1     1
  2      88     82.5    1     2
  3      85     88.0    2     2
  4             81.0    1     3
  5             86.5    2     3
  6             85.0    3     3
```

Also see WDIFF, WSLOPE, WINTERVAL, and WTEST.

WDIFF

Computes all possible differences between pairs of row elements from two columns.

Syntax

WDIFF   for C and C, put into C [put indices in C and C]

How to Use

This command, useful for nonparametric tests and confidence intervals, calculates all differences of the form $X_i - Y_j$ for X values from the first column and Y values from the second. Differences are stored in the third column listed. If two additional columns are specified, they will contain the row numbers i and j used for each difference, as shown in Fig. 6–79.

FIGURE 6-79

```
STAT> RETRIEVE 'EXAMPLE'
STAT> WDIFF C8 C9 C10 C11 C12
STAT> PRINT C8-C12

ROW    C8    C9    C10    C11    C12

 1      1     5    -4      1      1
 2      2     6    -5      1      2
 3      3          -3      2      1
 4      4          -4      2      2
 5                 -2      3      1
 6                 -3      3      2
 7                 -1      4      1
 8                 -2      4      2
```

Also see WALSH and MANN-WHITNEY.

WIDTH

Controls the width of plots.

Syntax

WIDTH of plots that follow is **K** spaces

How to Use

You can set the width of all subsequent BOXPLOT, DOTPLOT, LPLOT, MPLOT, PLOT, and TPLOT (but not TSPLOT) displays with this command. WIDTH may be any number from 10 to 150; its default value is 57 spaces.

Also see HEIGHT, OW, and OH.

WINTERVAL

Calculates a one-sample Wilcoxon rank estimate and confidence interval for the center of a symmetric population

Syntax

WINTERVAL [K% confidence] for data in C . . . C

How to Use

This command is a nonparametric alternative to TINTERVAL. It assumes only that each column of data is a random sample from a continuous, symmetric population. STAT101 calculates an approximate 95% confidence interval if you do not enter a value for K (which may be expressed as a percentage or as a decimal, such as 0.95).

FIGURE 6-80

```
STAT> RETRIEVE 'STA261'
STAT> WINTERVAL 'FINAL'

              ESTIMATED    ACHIEVED
          N    MEDIAN     CONFIDENCE   CONFIDENCE INTERVAL
FINAL    45     74.50        95.0    (  69.00,    78.50)
```

The estimated median shown in Fig. 6-80 is the median of the Walsh averages for the column. The confidence interval for the population median is based on the Wilcoxon test (WTEST). It is displayed together with its achieved confidence level, which is as close to the value of K as possible.

Also see WALSH and WTEST.

WRITE

Writes data in the current worksheet to a file or to the screen.

Syntax

WRITE [to 'FILENAME'] the data in columns C . . . C

FORMAT (format statement)

How to Use

The WRITE command saves data from your current worksheet in a standard format (ASCII) text file that can be edited, printed out, or read by packages other than STAT101. Only data from columns are written to the specified file; column names and constants are not. The default file extension for a file created by the WRITE command is .DAT, though you may specify your own extension. For example:

⇨ ```
STAT> RETRIEVE 'TREES'
STAT> WRITE 'TREES' C1-C3
```

stores the data from columns C1 through C3 of your current worksheet in a file TREES.DAT in your current directory. If no file name is specified, WRITE displays data on your screen.

Files created with WRITE may be input with the READ, SET, or INSERT commands. If you WRITE to a file and all the columns do not fit on one line, STAT101 puts the continuation symbol (&) at the end of the line and continues onto the next. If you WRITE columns of unequal length to a file, STAT101 adds missing value symbols (*) to the bottom of the shorter columns.

## Subcommand

**FORMAT**     (format statement)

The FORMAT subcommand tells STAT101 exactly where and how the data should appear on the output line. The format statement is always enclosed by parentheses. See online HELP WRITE FORMAT for more explanation and examples.

*Also see* SAVE, PRINT, READ, SET, and INSERT.

# WSLOPE

Computes slopes for all pairs of rows from two columns.

## Syntax

**WSLOPE**     y in C, x in C, put slopes in C [put indices in C,C]
```

How to Use

This nonparametric command is useful in finding robust estimates of the slope of a line through the data from two columns, with y-coordinates in column one and x-coordinates in column two. WSLOPE calculates the slopes of all lines through pairs of points (X_i, Y_i) and (X_j, Y_j) from rows i and j of the input columns and stores these slopes in the third column listed. You can also allocate two columns for the indices i and j, as shown in Fig. 6–81.

FIGURE 6–81

```
STAT> RETRIEVE 'EXAMPLE'
STAT> WSLOPE C8 C10-C13
STAT> PRINT C8 C10-C13

ROW   C8    C10       C11      C12   C13
  1    1    1.1     1.11111     2     1
  2    2    2.0     0.58824     3     1
  3    3    4.5     0.40000     3     2
  4    4    2.0     3.33333     4     1
  5                    *         4     2
  6               -0.40000       4     3
```

As shown in Fig. 6–81, if there are missing values, if the two input columns are of unequal length, or if a slope is undefined (e.g., the slope of a vertical line), the slope is stored as a missing value (∗).

WTEST

Does a one-sample Wilcoxon signed-rank test for one or more columns.

Syntax

| **WTEST** | [of center K] on the data **C** ... **C** |
| **ALTERNATIVE** | = K |

How to Use

WTEST is a nonparametric alternative to TTEST, assuming only that each column of data is a random sample from a symmetric population. It performs a two-sided test of whether the center (median) is equal to the number K. You can also use the ALTERNATIVE subcommand to do a one-sided test. The default value of K is 0.

FIGURE 6-82

```
STAT> RETRIEVE 'STA261'
STAT> WTEST 75 'FINAL'

TEST OF MEDIAN = 75.00 VERSUS MEDIAN N.E. 75.00

             N FOR   WILCOXON            ESTIMATED
         N   TEST    STATISTIC  P-VALUE  MEDIAN
FINAL   45   45      497.0      0.821    74.50
```

In Fig. 6-82, N FOR TEST is the number of observations not equal to K. Walsh averages are calculated for these values. The Wilcoxon statistic is the number of Walsh averages greater than K, plus one-half the number of Walsh averages equal to K. (The Wilcoxon statistic can also be computed in terms of signed ranks.)

The p-value is the probability, assuming the null hypothesis is true, of a Wilcoxon statistic being as extreme as, or more extreme than, the one observed. The estimated median shown is the median of the Walsh averages.

Subcommand

ALTERNATIVE = K

Performs a one-sided test of the alternative hypothesis H1:(center < K) for ALTERNATIVE = -1, or H1:(center > K) for ALTERNATIVE = 1. The null hypothesis is H0:(center = K) for both cases. (ALTERNATIVE = 0 does the same test as WTEST without the ALTERNATIVE subcommand.)

Also see WALSH and WINTERVAL.

ZINTERVAL

Calculates a confidence interval for the mean of each column.

Syntax

ZINTERVAL   [K% confidence] assumed σ = K, for data in C . . . C

How to Use

With this command you can calculate a confidence interval for the mean of one or more variables with population standard deviation σ = K. The default confidence level is 95%. See Fig. 6–83 for an example.

FIGURE 6-83

```
STAT> RETRIEVE 'STA261'
STAT> ZINTERVAL 15 'FINAL'

THE ASSUMED SIGMA =15.0
             N      MEAN    STDEV   SE MEAN    95.0 PERCENT C.I.
FINAL       45     72.18    18.19    2.24    ( 67.79,   76.57)
```

If you specify a fractional value for confidence K, say 0.80, STAT101 automatically multiplies it by 100 to give 80% confidence.

Also see ZTEST, TINTERVAL, TTEST, WINTERVAL, and SINTERVAL.

ZTEST

Performs a z-test on one or more columns.

Syntax

ZTEST   [of μ = K] assumed σ = K on C . . . C

ALTERNATIVE   = K

How to Use

This command does a two-sided test on each column of data, computing the appropriate p-value. You can also use the ALTERNATIVE subcommand to do a one-sided test. The default value for μ is 0. See Fig. 6–84 for an example.

FIGURE 6–84

```
STAT> SET C1
DATA>   40.1  30.2  39.4  39.8  39.0
DATA> END
STAT> ZTEST 40 0.4 C1

TEST OF MU = 40.000 VS MU N.E. 40.000
THE ASSUMED SIGMA = 0.400

              N      MEAN     STDEV    SE MEAN         Z    P VALUE
C1            5    37.700     4.213      0.179    -12.86     0.0000
```

Appendix A: Troubleshooting

STAT101 catches most mistakes when you make them: just after you press [Enter] after typing a command. It does not detect mistakes in the Data Editor until it attempts to execute commands on the data.

If you have trouble using STAT101, look for help in the following places:

- This user's manual
- Online help, either from the Session screen or Data Editor (the Data Editor has a special Troubleshooting topic)
- This appendix, Troubleshooting
- The file **READ.ME** that was installed with STAT101

The READ.ME file contains a listing of known STAT101 problems and includes strategies for working around them. To view this file from the DOS prompt, use any word processor or text editor, or

- Type **CD \STAT101** at the DOS prompt.
- Type **MORE< READ.ME** at the DOS prompt.

The DOS MORE command displays the READ.ME text file one screen at a time. When you have read one screen:

- Press [Enter] to see the next screen, and repeat until you are finished viewing the file. DOS returns you to the C:\STAT101> prompt.

This is an explanation of each problem, followed by a cause and corrective action.

Error detection and messages are safeguards used by STAT101 to help users avoid mistakes that could produce incorrect answers or destroy valuable data. This section explains common error conditions and provides corrective actions.

Not Reading Drive

⇨ Not ready reading drive A
   Abort, Retry, Fail?

Cause: The diskette drive specified by a STAT101 command is empty, the drive latch is not closed, the diskette is placed in the wrong direction, or the drive is faulty.

Action: Remove the diskette and insert the correct one in the specified drive and close the latch. Press [F] for Fail to return to STAT101.

Use Abort and Retry with caution. Abort exits STAT101 without giving you the chance to save your worksheet, while under certain conditions Retry can destroy the files on the diskette by overwriting the directory.

Can't Find File

⇨ * ERROR * Requested file does not exist

Cause: STAT101 cannot locate the file requested by RETRIEVE, READ, SET, INSERT, or EXECUTE. Either the file is not in the directory you specified or the command you are using is not appropriate for the file type.

Action: Include the location of the file (the path) within the single quotes that contain the file name. For example, use the command

⇨ STAT> RETRIEVE 'C:\STAT101\DATA\PULSE'

If you still get the same error message, it could be that the file you want to open is incompatible with the command you are using. Use the READ, SET, or INSERT commands with ASCII (.DAT) files, and the RETRIEVE command with saved worksheet (.MTW) files.

For more information on file types, use the command

⇨ STAT> HELP OVERVIEW 7

File Name Problem

⇨ * Check for filename with no quote
* ERROR * Missing or stray quote
* Name or file may be ignored
* Check for filename with no quote

Cause: One of these messages is displayed when you forget to enclose the file name in single quotes, or you use the wrong type of quotes.

Action: Enter the STAT101 command with the file name, including path information, correctly enclosed in single quotes. For example, use the command

⇨ `STAT> RETRIEVE 'C:\STAT101\DATA\PRES'`

Can't Start STAT101

⇨ `Bad command or filename`

Cause: DOS cannot locate the command file used to start STAT101.

Action: The directory that contains STAT101 must be included in the prompt. When you installed STAT101, you were given the opportunity to create a directory on your hard disk called STAT101. If you did so, you can use the commands

⇨ `C:\> CD \STAT101`
`C:\STAT101> STAT101`

This tells DOS to change to the STAT101 directory and look there for the command file (STAT101.EXE). If you installed STAT101 in a different directory, use the CD command (Change Directory) to enter that path instead. You can also add STAT101.EXE to your AUTOEXEC.BAT file, in which case you can reboot and start STAT101 at any prompt by just typing

⇨ `C:\> STAT101`

Appendix B: Sample Data Sets

This manual uses a variety of data sets to illustrate how STAT101 commands operate on real-life data. All of these data sets come with STAT101, and were automatically placed in the STAT101 directory when you installed STAT101. This appendix lists and describes all data sets included on the program disk.

To open one of the sample data sets, use the RETRIEVE command in the Session screen with the file name in single quotes. For example, to open POTATO and see what it contains:

- Type **RETRIEVE 'POTATO'** after the STAT> prompt and then press Enter.
- Type **INFO** and press Enter to get worksheet information.
- Press Esc to view the data set in the Data Editor.

See Retrieving Sample Data Sets in Chapter 5, Handling Files and Printing, for more information.

1STGRADE

As part of a class exercise, a first-grade teacher recorded the heights and weights of his 73 students, hoping to determine how height and weight were related, and whether the class was normal with respect to U.S. government standards on height and weight.

Column Name	Count	Description
1 HGT	73	Heights of students in inches
2 WGT	73	Weights of students in pounds

ACCOUNTS

As branch manager of Fly-By-Night International Bank's largest branch, you want to see your customers served quickly, especially when they are opening a new account. You would like it to take no more than six minutes to fill out and process all of the paperwork to open any kind of new account.

Your assistant manager measures the time (in minutes) it takes to process five randomly selected "new account" customers each day for 40 consecutive days.

Column Name	Count	Description
1 CUST1	40	Time for the first randomly selected customer on a given day to be processed
2 CUST2	40	Time for the second randomly selected customer on a given day to be processed
3 CUST3	40	Time for the third randomly selected customer on a given day to be processed
4 CUST4	40	Time for the fourth randomly selected customer on a given day to be processed
5 CUST5	40	Time for the fifth randomly selected customer on a given day to be processed

ACID

A chemistry class recorded titration results from two experiments. This data set contains both experiments; while ACID1 contains the first and ACID2 the second.

Column Name	Count	Description
1 ACID1	124	Titration results
2 ACID2	37	Titration results

ACID1

This is the first experiment described in ACID.

Column Name	Count	Description
1 ACID1	124	Titration results

ACID2

This is the second experiment described in ACID.

Column Name	Count	Description
2 ACID2	37	Titration results

AGGRESS

In a study of cadmium, researchers exposed twelve fish to cadmium. See CD1 and CD2 for an introduction to the cadmium investigation. In this part of the study, they recorded the aggressive behavior of 10 fish (5 controls and 5 exposed to cadmium), both before and after exposure, and the number of aggressive encounters, averaged over a week.

Column Name	Count	Description
1 CONT BFR	5	Encounters for control fish before exposure to tap water
2 CONT AFT	5	Encounters for control fish after a week of exposure to tap water
3 CD BFR	5	Encounters for cadmium fish before exposure to cadmium
4 CD AFT	5	Encounters for cadmium fish after a week of exposure to cadmium

ALFALFA

A University of Wisconsin researcher tested the yields of six varieties of alfalfa on each of four separate fields.

Column Name	Count	Description
1 YIELD	24	The yield for this particular variety
2 VARIETY	24	Variety number (1-6)
3 FIELD	24	Field number (1-4)

AZALEA

This experiment tested the effects of ozone on eleven varieties of azalea plants (just two are included here, labeled A and B). New plants were used each week, and environmental conditions differed from week to week.

Column Name	Count	Description
1 A-WEEK-1	10	Leaf damage for Week 1, variety A
2 A-WEEK-2	10	Leaf damage for Week 2, variety A
3 A-WEEK-3	10	Leaf damage for Week 3, variety A
4 A-WEEK-4	10	Leaf damage for Week 4, variety A
5 A-WEEK-5	10	Leaf damage for Week 5, variety A
6 B-WEEK-1	10	Leaf damage for Week 1, variety B
7 B-WEEK-2	10	Leaf damage for Week 2, variety B
8 B-WEEK-3	10	Leaf damage for Week 3, variety B
9 B-WEEK-4	10	Leaf damage for Week 4, variety B
10 B-WEEK-5	10	Leaf damage for Week 5, variety B

BANKING

Bank examiners obtained banking profits for the first quarter of the year for all banks in a state in the years 1991 and 1990. They chose ten states randomly each year and obtained the profits from FDIC records. The total first-quarter profits for each of the 10 states are stored in separate columns, with each year's data in one column.

Column Name	Count	Description
1 1991	10	First-quarter 1991 banking profits in billions for all banks in 10 randomly selected states
2 1992	10	First-quarter 1990 banking profits in billions for all banks in 10 randomly selected states

BPRES

Researchers interviewed nine subjects to investigate the relationship among blood pressure, diet, and fitness.

Column Name	Count	Description
1 BP	9	Number of points above normal diastolic blood pressure
2 OVER WGT	9	Number of kilograms overweight
3 FATS	9	Average number of grams saturated fatty foods consumed per day
4 EXERCISE	9	Average number of minutes of exercise per day

CAMSHAFT

An automobile assembly plant with four shifts, concerned about quality control, measured sets of five camshafts each shift for a total of twenty samples daily. The column Length contains measurements from all the camshafts used at the plant, while Supp1 and Supp2 are measurements of shafts from Supplier 1 and Supplier 2, respectively.

Column Name	Count	Description
1 Length	100	Camshaft length in mm
2 Supp 1	100	Lengths of camshafts from Supplier 1
3 Supp 2	100	Lengths of camshafts from Supplier 2

CARSPEED

(From *Statistics Today*, Byrkit, p. 634)

Researchers monitored speed along a certain highway, using four different speed control methods, including various speed limit signs and a parked state highway patrol car. They obtained speeds of out-of-state cars for each speed control method. The data set SPEED2 contains this same data with the speeds in one column and method identifiers in the other.

Column Name	Count	Description
1 METHOD 1	12	Speeds of cars using method 1
2 METHOD 2	10	Speeds of cars using method 2
3 METHOD 3	11	Speeds of cars using method 3
4 METHOD 4	11	Speeds of cars using method 4

CARTOON

Educators who use instructional films wanted to evaluate the relative effectiveness of different visual aids to see how well people watching the films learn and retain the material. One hundred seventy-nine participants attended a short lecture on how people take on different roles in groups. The lecturer showed 18 slides that identified each group role with an animal, shown once in a cartoon sketch and once in a realistic picture. All participants saw all of the 18 slides, but one group (randomly selected) saw them in color while the other group saw them in black and white.

Immediately after the lecture, the participants were shown the slides again in random order. They were told to write down the group role represented by the animal in each slide. The investigators tallied and recorded the number of cartoon characters and realistic characters that the participants correctly identified. Four weeks later, the participants took the same test and their scores were computed again (those that didn't show up were entered as missing values).

The investigators gave all participants the OTIS Quick Scoring Mental Ability Test to get a rough measure of each participant's natural ability. The participants were preprofessional and professional personnel at three Pennsylvania hospitals involved in an in-service training program, as well as a group of Penn State undergraduates.

Column Name	Count	Description
1 ID	179	Identification number
2 COLOR	179	0 = Black and white 1 = Color

3 ED	179	Education: 0 = Preprofessional 1 = Professional 2 = College student
4 LOCATION	179	Location: 1 = Hospital A 2 = Hospital B 3 = Hospital C 4 = Penn State student
5 OTIS	179	OTIS Score: from between 70 to 130
6 CARTOON 1	179	Cartoon test score (immediately after viewing): from 0 to 9
7 REAL1	179	Realistic test score (immediately after viewing): from 0 to 9
8 CARTOON2	179	Cartoon test score (4 weeks later): from 0 to 9; * is missing data
9 REAL2	179	Realistic test score (4 weeks later): from 0 to 9; * is missing data

CD1

In order to study cadmium, researchers exposed twelve fish to cadmium. After exposure, they measured cadmium concentration in the gill, liver, and kidney. They also studied five control fish, and after a similar time in ordinary water, measured their gill, liver, and kidney concentrations as well. See CD2 and AGGRESS for other aspects of this study.

Column Name	Count	Missing	Description
1 CTL FISH	5		Control fish number
2 CTL LIV	5		Control fish liver cadmium concentration
3 CTL KIDN	5		Control fish kidney cadmium concentration *(continued)*

4 CTL GILL	5	1	Control fish gill cadmium concentration
5 CD FISH	12		Cadmium fish number
6 CD LIV	12		Cadmium fish liver cadmium concentration
7 CD KIDN	12		Cadmium fish kidney cadmium concentration
8 CD GILL	12	1	Control fish gill cadmium concentration

CD2

This data set contains the same information as CD1, but the data have been stacked into 5 columns instead of 8.

Column Name	Count	Missing	Description
1 FISH	17		Fish number
2 LIVER	17		Liver concentrations
3 KIDNEY	17		Kidney concentrations
4 GILL	17	2	Gill concentrations
5 GROUP	17		Group number (1=control, 2=cadmium)

CHOLEST

Medical researchers recorded blood cholesterol levels of 28 heart-attack victims 2, 4, and 14 days following the attack. The levels of 30 individuals who had not had an attack were taken as a control. CHOLESTC and CHOLESTE contain subsets of this data set.

Column Name	Count	Missing	Description
1 2-DAY	28		Cholesterol level 2 days after heart attack
2 4-DAY	28		Cholesterol level 4 days after heart attack
3 14-DAY	28	9	Cholesterol level 14 days after heart attack
4 CONTROL	30		Cholesterol levels of control group

CHOLESTC

This is a subset of the CHOLEST data set: the control group data only.

CHOLESTE

This is a subset of the CHOLEST data set: the experimental group data only.

CITIES

Monthly average temperatures for five U.S. cities were recorded for each of 12 months.

Column Name	Count	Description
1 MONTH	12	Month
2 ATLANTA	12	Average temperatures in Atlanta
3 BISMARK	12	Average temperatures in Bismarck
4 NEW YORK	12	Average temperatures in New York
5 SAN DIEG	12	Average temperatures in San Diego
6 PHOENIX	12	Average temperatures in Phoenix

CLOTTING

Aspirin is routinely prescribed by doctors to heart-attack patients to prevent the formation of obstructions in the arteries due to blood clots. A study was done to investigate whether the formation of blood clots is affected by aspirin. Twelve adult men were used in the study. The time for these twelve men's blood to clot was measured by the time from the start of a prothrombin-thrombin reaction to the formation of the final clot. The clotting time was measured before the men took two aspirin and then three hours after taking the aspirin.

Column Name	Count	Description
1 BEFORE	12	The time for a clot to form, before taking aspirin
2 AFTER	12	The time for a clot to form, after taking aspirin

COMPUTER

After saving up some money from various summer jobs, a student who wanted to buy a computer began to research prices and equipment. She decided to purchase a high-performance computer, so that it would be usable throughout her college years and well beyond. The college recommends that students purchase DOS personal computers, so her efforts were directed at investigating these kinds of machines. To research the cost, she bought several personal computer magazines, and recorded the prices and features of various computers advertised.

Column Name	Count	Description
1 CPU	35	The type of CPU used in the computer
2 SPEED	35	The speed of the CPU
3 HD	35	The size of the hard disk drive in megabytes
4 FLOPPY	35	The number of floppy drives included
5 MEMORY	35	The RAM size in megabytes
6 MONITOR	35	Color or monochrome display

7 MOUSE	35	Is a mouse included?
8 SOFTWARE	35	Is software included?
9 PRICE	35	The cost of the system in $

COREDATA

A geology graduate student obtained core samples from several locations; this data set was taken from one of those samples. The student measured the percent mud in the sample, approximately every 10 feet.

Column Name	Count	Description
1 DEPTH	21	Depth measured in feet
2 % MUD	21	Percent mud at the depth

CRANKSH

In an operating engine, parts of the crankshaft move up and down. AtoBDist is the distance (in mm) from the actual (A) position of a point on the crankshaft to a baseline (B) position. To ensure production quality, a manager took five measurements each working day in a car assembly plant, from September 28 through October 15, and then ten per day from the 18th through the 25th.

Column Name	Count	Description
1 AtoBDist	125	Distance from A to B in mm
2 Month	125	Month measurements were taken
3 Day	125	Day measurements were taken

CROWDS

In a study of perceived crowdedness and the color of the walls of a room, an experimental psychology professor selected 36 unacquainted female and 36 unacquainted male subjects. The female subjects were all brought into a room in which the walls had been painted a dark green. At the end of one hour each subject was asked to rate how crowded the subject felt on a scale of 1 to 30, with 30 being the most crowded. The experiment was repeated with the male subjects.

The perceived crowdedness of each female subject is recorded in column C1 and the male values are recorded in column C2.

Column	Count	Description
1 FEMRATE	36	The perceived crowdedness of females in the study
2 MALERATE	36	The perceived crowdedness of males in the study

EMPLOY

This file contains numbers of employees in three industries in Wisconsin, Wholesale and Retail Trade, Food and Kindred Products, and Fabricated Metals, measured each month over five years.

Column Name	Count	Description
1 TRADE	60	Employees in thousands
2 FOOD	60	Employees in thousands
3 METALS	60	Employees in thousands

EXAM

These data are the test scores (out of 100) of two hourly exams given in a small graduate statistics class.

Column Name	Count	Description
1 EXAM 1	8	Exam 1 results
2 EXAM 2	8	Exam 2 results

EXAMPLE

This is fabricated data designed to exemplify certain commands.

Column Name	Count	Missing	Description
1 SEX	7		Subject's sex
2 DIET	7		Whether or not subject was dieting
3 HT	7		Height
4 WT	7	2	Weight
5 CHANGE	7	2	Miscellaneous
6		2	Miscellaneous
7		2	Miscellaneous
8		4	Miscellaneous
9		2	Miscellaneous
10		4	Miscellaneous
11		4	Miscellaneous

FA

This data set was generated to make a point about regression.

FABRIC

The fire department laboratory tested the flammability of fabric, using the same methods in five different labs. The measurements are the length of the burned portion of a piece of fabric held over flame for a fixed amount of time.

Column (not named)	Count	Description
1	11	Length of charred fabric
2	11	Length of charred fabric

(continued)

3	11	Length of charred fabric
4	11	Length of charred fabric
5	11	Length of charred fabric

FALCON

A study to investigate the residual effect of the pesticide DDT in falcons involved capturing and measuring DDT in falcons from three nesting areas (US, Canada, and the Arctic region) and three different age groups (young, middle-aged, and old).

Column Name	Count	Description
1 DDT	27	The DDT measurement
2 SITE	27	The nesting site of the bird: 1 = US 2 = Canada 3 = Arctic region
3 AGE	27	The age of the bird: 1 = young 2 = middle-aged 3 = old

FISH

Scientists are concerned with the effects of acid rain on aquatic life. Studies have shown that at very low pH levels (around pH 3, which is very acidic), many fish die. In the environment, however, the decrease in pH from naturally occurring levels (of around 7) is rather gradual and rarely are very low pH values seen. Scientists are interested in the effects of this lowered pH on fish, particularly in how the lowered pH adversely changes the parental behavior in fish. Fish offspring that are not cared for adequately have a lower survival rate.

In one study, they observed several parental behaviors of a certain species of fish. Guarding the nest is an important activity because it ensures that the fish eggs and young are safe from predators. Fanning the eggs is also important since it keeps oxygenated water circulating over the eggs and hence increases the chances of survival. Pairs of fish were exposed to water of varying pH levels and the percentage of time each fish spent guarding and fanning its young was recorded. The values given are the averages of approximately 10 fish.

Column Name	Count	Description
1 pH	80	The level of pH the fish were exposed to: 7.0, 6.5, 6.0, and 5.5
2 DAY	80	The day of the study, 1 to 10 days
3 GENDER	80	The gender of the fish, M or F
4 GUARDING	80	The average time spent guarding the young
5 FANNING	80	The average time spent fanning the young

FLEXIBLE

Flexible Circuits, Inc. is a company specializing in the manufacture of parts used in various electronic components. Their products include copper wires used in stereo equipment. These copper wires are subjected to a plating process in which they are dipped in a micro-etch bath, set at 95° F, for 1 minute and 10 seconds. The diameters of the wires must meet certain specifications if they are to be acceptable to the purchaser.

Two studies investigated the apparent lack of consistency of the wire diameters from shift to shift. In the first study, the temperature of the bath was set at 95° F and wires were randomly selected from each shift's production: nine from the first shift, six from the second, and three from the third shift.

In the second study, the temperature of the bath was set to 100° F. Twenty-five samples were taken: twelve from shift 1, seven from shift 2, and six from shift 3.

Column	Count	Description
1 DIAMETER	18	The diameter in microinches of wires used in stereo equipment
2 SHIFT	18	The shift during which the wire was produced

(continued)

3	DIAM1	12	Wire diameters in microinches taken from shift 1
4	DIAM2	7	Wire diameters in microinches taken from shift 2
5	DIAM3	6	Wire diameters in microinches taken from shift 3

FLEX2

In another study of the diameters of the copper-plated wires manufactured by the Flexible Circuits, Inc., the stability of the production process was investigated. The company's engineers had specified wire diameters of 70 microinches, ±3 microinches. Five wires were randomly selected each hour for 21 consecutive hours.

Several weeks later, after instituting new guidelines for the production process, wire diameters were selected systematically every 12 minutes during each of the three shifts' production on a given day. These data are stored in columns SHIFT 1, SHIFT 2, and SHIFT 3, respectively.

As a final study of the wire production process, the plant engineer wanted to determine what percentage of the wires produced were defective (did not meet the specifications set forth by the engineering staff). Two hundred wires were randomly selected each day for an entire month. C5 contains the number of nonconforming wires out of 200.

Column Name	Count	Description
1 DIAMETER	105	The diameter, sampled 5 per hour for 21 hours
2 SHIFT1	40	Wire diameter, systematically selected every 12 minutes from shift 1
3 SHIFT2	40	Wire diameter, systematically selected every 12 minutes from shift 2
4 SHIFT3	40	Wire diameter, systematically selected every 12 minutes from shift 3
5 NONCONF	30	The number of defective (nonconforming) wires out of 200 wires selected randomly each day for a month

FOOTBALL

The data below come from a newspaper article listing the salaries of all players on a professional football team prior to the start of the 1991 season.

Column Name	Count	Missing	Description
1 POSITION	50		The position of the player
2 EXPERNCE	50		The years' experience of the player
3 1990 SAL	47	3	The 1990 salary of the player
4 1991 SAL	39	11	The 1991 salary of the player
5 1992 SAL	19	31	The 1992 salary of the player
6 BONUS	17	33	The signing bonus the player received

FURNACE

Wisconsin Power and Light measured energy consumption for 90 gas-heated homes to determine the relative efficiency of two venting systems: electric vent dampers (EVD) that close the vent when the furnace is in its resting cycle, and thermally activated vent dampers (TVD) that close the vent according to the thermal properties of a set of bimetal fins in the vent. The measurements were taken over a period of several weeks, including times when the vents were operating and when they were not.

To adjust for house size and weather conditions, the power company used a simple formula to record average energy used by each house: (consumption in BTUs)/[(weather in degree days)(house area in square feet)]. They also recorded relevant characteristics of each house and chimney.

Column Name	Count	Missing	Description
1 TYPE	90		Furnace type: 1 = forced air 2 = gravity 3 = forced water
2 CH.AREA	90	1	Area of chimney
3 CH.SHAPE	90	1	Shape of chimney: 1 = round 2 = square 3 = rectangular
4 CH.HT	90		Height of chimney, in feet
5 CH.LINER	90	1	Chimney liner: 0 = unlined 1 = tile 2 = metal
6 HOUSE	90		House type: 1 = ranch 2 = two-story 3 = tri-level 4 = bi-level 5 = one and a half stories
7 AGE	90		Age of house (for houses 100 years or older, 99 was entered)
8 BTU.IN	90		Average energy consumption with damper in
9 BTU.OUT	90		Average energy consumption with damper out

10 DAMPER		90	Damper type: 1 = EVD 2 = TVD

GENETICS

Scientists wanted to determine whether carbon-dioxide-deficient soil changes the distribution of phenotypes of peas. Two hundred randomly chosen peas were grown in carbon-dioxide-deficient soil and the phenotype of each of these peas was determined.

Column Name	Count	Description
1 TYPE	4	Phenotype of pea: 1 = smooth-yellow 2 = smooth-green 3 = wrinkled-yellow 4 = wrinkled-green
2 PROB	4	Mendelian probabilities of each phenotype
3 PHENO	200	Phenotype code for each of 200 randomly chosen peas grown in carbon-dioxide-deficient soil

GOLF

A college student, who is also an avid golfer, kept track of his golf scores in the fall of 1991. Based on playing nine holes, he recorded his score, how many greens he hit in regulation, and how many pars or better he made for the nine holes.

Column Name	Count	Description
1 SCORE	22	The score for nine holes of golf
2 REGLTN	22	The number of greens made in the regulation number of strokes
3 PARS	22	The number of pars or better (pars, birdies, eagles, etc.) made in the nine holes

GRADES

Investigators collected data to study the relationship between SAT scores (Scholastic Aptitude Tests, often used as college admissions or course placement criteria) and grade point averages (GPA on a 0.0 to 4.0 scale). They sampled 200 students from a northeastern university. GRADES contains the GPA and verbal and mathematical SAT scores for all 200 students.

Column Name	Count	Description
1 VERBAL	200	Verbal aptitude test scores
2 MATH	200	Mathematics aptitude test scores
3 GPA	200	Grade point average, from 0 to 4

GB

The GRADES data were broken down into two random samples of 100 each for ease of use. GB contains one of the samples.

GA

The GRADES data were broken down into two random samples of 100 each for ease of use. GA contains one of the samples.

GUARD

A study investigating the effect of water acidity/alkalinity (pH) on the parental care behavior of fish involved forty pairs of fish, male and female. Each pair was housed in individual aquaria and exposed to one of four levels of pH: 7.0, 6.5, 6.0, and 5.5 (10 pairs per level). The investigators monitored the fish daily for ten days. The data represent the average time spent guarding the nest for each gender on each day for each of the 4 pH groups. In each column the first 10 observations are the average of the male fish for days 1-10 for that exposure group and the second 10 observations are the average for the female fish for days 1-10 for that exposure group.

Column Name	Count	Description
1 PH 7.0	20	Data for pH 7.0
2 PH 6.5	20	Data for pH 6.5

3 PH 6.0	20	Data for pH 6.0
4 PH 5.5	20	Data for pH 5.5

HCC

A Wisconsin psychiatric health care center kept records on the month each patient was admitted, the length of stay, and the reason for discharge, either a 1 for normal or 2 for other (against medical advice, court-ordered, absent without leave, etc.).

Column Name	Count	Description
1 REASON	58	Reason for leaving: 1 = Normal 2 = Other
2 MONTH	58	Month admitted
3 LENGTH	58	Length of stay, in days

HEALTH

As part of a study investigating the general health of college students, information was collected from 21 randomly chosen college students.

Column Name	Count	Description
1 AGE	21	The age of the student in years
2 GENDER	21	The gender of the student: 0 = Male 1 = Female
3 HGT	21	The height of the student in centimeters
4 DBP	21	The diastolic blood pressure of the student
5 SBP	21	The systolic blood pressure of the student

(continued)

6 SMOKER?	21	The smoking status of the student: 0 = No 1 = Yes	
7 EX REG	21	The volume of air exhaled after a regular breath	
8 EX MAX	21	The volume of air exhaled after one deep breath	

HEART

The single column in this data set contains the age of twenty male heart transplant patients at the time of their first heart transplant.

Column Name	Count	Description
1 AGE	20	Age at first heart transplant

HEIGHTS

A professor in the Physiology department of a medical school collected data on the self-reported height of 60 females in an introductory physiology class. Students were asked to provide their height (in inches) the first day of class.

Column Name	Count	Description
1 ID	60	The student's ID #
2 HEIGHTS	60	The self-reported heights (in inches of 60 college females)

HOMESALE

The single column of this data set contains the number of homes sold by a local realtor over the past four years for each month.

Column Name	Count	Description
1 SOLD	47	Number of homes sold per month

HWASTE

New and abandoned hazardous waste sites are being discovered across the U.S. This data set lists the number of hazardous waste sites found in each region of the country.

Column Name	Count	Description
1 NORTHWST	7	The number of hazardous waste sites found in each state in the northwestern part of the U.S.
2 SOUTHWST	6	The number of hazardous waste sites found in each state in the southwestern part of the U.S.
3 MIDWEST	7	The number of hazardous waste sites found in each state in the midwestern part of the U.S.
4 SOUTHCEN	7	The number of hazardous waste sites found in each state in the south-central part of the U.S.
5 SOUTHEAS	12	The number of hazardous waste sites found in each state in the southeastern part of the U.S.
6 NORTHEAS	9	The number of hazardous waste sites found in each state in the northeastern part of the U.S.

KRUNCHY

The marketing department of Krunchy Cereal Company did a study to determine which of two advertising companies (Ace or Best) to use to advertise their product lines. Each of the two companies uses three different advertising media in their ad campaigns: newspaper, radio, and TV. On a trial basis both companies and all three ad campaign types were used in a preliminary study. Twelve comparable cities were randomly selected and two cities were used for each company and media type combination. The sales increase per advertising dollar was computed after each ad campaign.

Column Name	Count	Description
1 SALESINC	12	Sales increase per advertising dollar for a city
2 MEDIUM	12	The medium used in the advertising campaign: 1=newspaper 2=radio 3=TV
3 AGENCY	12	The ad agency used in the campaign: 1=Ace 2=Best

LAKE

This data set contains characteristic information on the lakes in Vilas and Oneida counties in northern Wisconsin from the years 1959-1963.

Column Name	Count	Description
1 AREA	71	Area of lake in acres
2 DEPTH	71	Maximum depth of lake in feet
3 PH	71	pH, a measure of acidity
4 WSHED	71	Watershed area in square miles
5 HIONS	71	Concentration of hydrogen ions (directly related to pH)

LOTTO

A state lotto randomly selects six numbers from the numbers 1 through 47. There are two drawings per week. To check whether the drawings are done in a random fashion, a student monitored the game for several years. Based on information provided by the local newspaper, the student determined the number of times each of the numbers 1 through 47 had been drawn over the course of the past three years. Hence there were a total of 156 drawings.

Column Name	Count	Description
1 OBSERVED	47	The observed number of times each number has been drawn

MAPLE

In autumn, small winged fruit called samara fall off maple trees, spinning as they go. A forest scientist studied the relationship between how fast they fell and their "disk loading" (a quantity based on their size and weight). The samara disk loading is related to the aerodynamics of helicopters.

Column Name	Count	Missing	Description
1 LOAD1	12		Disk loading (from tree 1)
2 VEL1	12		Velocity (from tree 1)
3 LOAD2	12	1	Disk loading (from tree 2)
4 VEL2	12	1	Velocity (from tree 2)
5 LOAD3	12		Disk loading (from tree 3)
6 VEL3	12		Velocity (from tree 3)

MEATLOAF

This data set records the drip loss (liquid loss during cooking divided by original weight) of meatloaf cooked at eight different positions in an oven. There were three batches with eight loaves in each.

Column Name	Count	Description
1 DRIPLOSS	24	Drip loss quotient
2 BATCH	24	First, second, or third batch
3 POSITION	24	Position in oven

MIAMI

A professor at Miami University in Oxford, Ohio distributed a questionnaire to all females in an introductory statistics course to obtain demographic information. The data set contains the responses of 83 of the 197 females who participated in the study. This same data for the 83 females is stored as an ASCII text file in MIAMI.DAT.

Column Name	Count	Description
1 HAIR	83	Hair color: 1 = brown 2 = black 3 = blonde 4 = red
2 EYE	83	Eye color: 1 = brown 2 = blue 3 = green
3 SCHOOL	83	School of major: 1 = Applied Science 2 = Arts and Sciences 3 = Education 4 = Fine Arts 5 = Business School
4 SORORITY	83	Belong to sorority? 1 = yes 2 = no
5 OH RES	83	Ohio resident? 1 = yes 2 = no
6 BIRTH	83	Birth month

7 RELIGION	83	Religion: 1 = Roman Catholic 2 = Protestant 3 = Jewish 4 = Other
8 BROTHERS	83	Number of brothers
9 SISTERS	83	Number of sisters
10 GLASSES	83	Wear glasses? 1=yes 2=no
11 HAIR LNG	83	Length of hair: 1 = short 2 = medium 3 = long
12 CREDITS	83	Number of credits currently enrolled for
13 LIVING	83	Living arrangements: 1 = dorm 2 = apartment 3 = commute
14 BOYFRND	83	Boyfriend? 1 = yes 2 = no
15 ALCOHOL	83	Drink alcohol? 0 = no 1 = once a month 2 = once a week 3 = more than once a week

(continued)

16 P EARS	83	Number of pierced ears:	
		0 = none	
		1 = single	
		2 = double	
		3 = triple	
17 SLEEP	83	Average number of hours of sleep per night	
18 SOAPS	83	Watch soap operas regularly?	
		1=yes	
		2=no	
19 SMOKE	83	Smoke regularly?	
		1 = yes	
		2 = no	
20 DIET	83	Currently dieting?	
		1 = yes	
		2 = no	
21 JOB	83	Currently working?	
		1 = yes	
		2 = no	
22 HGT	83	Height (in inches)	
23 WGT	83	Weight (in pounds)	
24 GPA	83	Grade point average (rounded up to nearest .5 on a 4.0 scale; 2.0 is 2.0 and below)	

NIELSEN

Each year the Nielsen company does an extensive survey of TV viewing habits of the U.S. population. Many of the surveys are done by age and gender groups because sponsors are very interested in how these variables relate to the programs people watch. (For example, see the VPVH data set.)

Column Name	Count	Missing	Description
1 YEAR	25		Year
2 TV STNS	25	8	The number of TV stations, both VHF and UHF
3 NUM TVS	25	8	The number of households in the U.S. with TVs in 1,000
4 CABLE	25	11	The number of cable stations operating
5 SUBSCBRS	25	9	The number of cable subscribers in 1,000
6 TIME	25	5	The average number of hours of viewing per TV household per week

PAPER

A large metropolitan newspaper buys its newsprint from a supplier in the northwest. A quality engineer inspects each roll of paper that is used by the newspaper. The engineer is especially concerned about small holes in the paper, because if there is a hole, the print is lost and this makes it difficult, if not impossible, to read an article or ad.

Each roll is sampled six times. The sampling involves taking 1.5- by 2-foot sections of the paper (this corresponds roughly to the final dimension of a page of the newspaper) and counting the number of holes (or flaws) larger than 1 millimeter in diameter found in the section.

Column Name	Count	Description
1 SAMPLE 1	35	The number of holes per 1.5- by 2-foot section of newspaper in the first sample
2 SAMPLE 2	35	The number of holes per 1.5- by 2-foot section of newspaper in the second sample
3 SAMPLE 3	35	The number of holes per 1.5- by 2-foot section of newspaper in the third sample

4 SAMPLE 4	35	The number of holes per 1.5- by 2-foot section of newspaper in the fourth sample
5 SAMPLE 5	35	The number of holes per 1.5- by 2-foot section of newspaper in the fifth sample
6 SAMPLE 6	35	The number of holes per 1.5- by 2-foot section of newspaper in the sixth sample

PASTA

The nutritionist at a local hospital investigated the different pasta brands that were available from the hospital's food supplier. Each brand was rated by several members of the hospital's nutrition department. The average rating for each brand of pasta was recorded, as well as the cost and calories per 10-oz. serving of the pasta.

Column Name	Count	Description
1 RATING	15	The rating of a particular brand of pasta
2 CENT/SER	15	The cost (in cents) per 10-oz. serving of the pasta
3 CAL/SER	15	The calories per 10-oz. serving of the pasta

PAY

This data set contains the salary information of the 11 salaried employees in the sales department of the Technitron company. This data set is part of the larger data set TECHN, which contains the salary information of all employees in four departments at Technitron.

Column Name	Count	Missing	Description
1 SALARY	11		The salary of an employee in the sales department
2 YRS EM	11		The number of years employed at Technitron
3 PRIOR YR	11	1	The number of prior years' experience

4 EDUC	11	1	Years of education after high school
6 ID	11		The company identification number for the employee
7 GENDER	11		The gender of the employee
8 GENDER N	11		The coded gender of the employee, where 0 = female and 1 = male

PERU

Anthropologists wanted to determine the long-term effects of altitude change on human blood pressure. They measured the blood pressures of a number of Peruvians native to the high Andes mountains who had since migrated to lower climes. Previous research suggested that migration of this kind might cause higher blood pressure at first, but over time blood pressure would decrease. The subjects were all males over 21, born at high altitudes, with parents born at high altitudes. The measurements included a number of characteristics to help measure obesity: skin-fold and other physical characteristics. Systolic and diastolic blood pressure are recorded separately; systolic is often a more sensitive indicator. Note that this is only a portion of the data collected.

Column Name	Count	Description
1 AGE	39	Age in years
2 YEARS	39	Years since migration
3 WEIGHT	39	Weight in kilograms
4 HEIGHT	39	Height in mm
5 CHIN	39	Chin skin fold in mm
6 FOREARM	39	Forearm skin fold in mm
7 CALF	39	Calf skin fold in mm
8 PULSE	39	Pulse in beats per minute

(continued)

| 9 SYSTOL | 39 | Systolic blood pressure |
| 10 DIASTOL | 39 | Diastolic blood pressure |

PLYWOOD

A U.S. Forest Products Research Laboratory tested the effects of certain variables in the production of plywood. Chucks inserted at each end spin the logs, and a saw blade then cuts off a thin layer. This study measured the torque that could be applied to the chucks before they spun out, under different conditions of log temperature, log diameter, and chuck penetration.

Column Name	Count	Description
1 DIAMETER	24	Log diameter, in inches
2 PENETRTN	24	Distance chuck was inserted into the log
3 TEMP	24	Temperature of log
4 TORQUE	24	Amount of torque applied before chuck slipped out

POPLAR1

Clones are genetically identical cells descended from the same individual. Researchers have identified a single poplar clone that yields fast-growing, hardy trees. These trees may one day serve as an alternative to conventional fuel as an energy resource.

Researchers at The Pennsylvania State University planted Poplar Clone 252 on two different sites—one, a rich site by a creek, and the other, a dry, sandy site on a ridge. They measured the diameter in centimeters, height in meters, and the dry weight of wood in kilograms for a sample of three-year-old trees so that they could determine the weight of a tree from its diameter and height measurements. See POPLAR2 for further data.

Column Name	Count	Description
1 Diameter	15	Diameter of tree in cm
2 Height	15	Height of tree in meters
3 Weight	15	Dry weight of wood in kilograms

POPLAR2

In an effort to determine how to maximize yield, the same researchers described in POPLAR1 designed an experiment to determine how two factors, Site and Treatment, influence the weight of four-year-old poplar clones. They applied four different treatments to the trees in Site 1 (the rich, moist soil by a creek) and Site 2 (the dry, sandy soil on a ridge). Treatment 1 was the control (no treatment), Treatment 2 was fertilizer, Treatment 3 was irrigation, and Treatment 4 was both fertilizer and irrigation. To account for a variety in weather conditions, they replicated the data by planting half the trees in Year 1 and the other half in Year 2.

Column Name	Count	Description
1 Site	298	Site: 1 = rich, moist soil near a creek 2 = dry, sandy soil on a ridge
2 Year	298	First or second year
3 Treatmnt	298	Treatment: 1 = none 2 = fertilizer 3 = irrigation 4 = fertilizer and irrigation
4 Diameter	298	Diameter of tree in cm
5 Height	298	Height of tree in meters
6 Weight	298	Dry weight of wood in kilograms
7 Age	298	Age of tree in years

POTATO

The University of Wisconsin conducted a study of potatoes in which rot-causing bacteria were injected into the potatoes in varying amounts. The potatoes were left for five days at various temperatures and in atmospheres with different oxygen contents. The diameter of the rotted area was measured in mm.

Column Name	Count	Description
1 BACTERIA	54	Amount of bacteria injected into the potato: 1 = Low 2 = Medium 3 = High
2 TEMP	54	Temperature during storage: 1 = 10 ° Celsius 2 = 16 ° Celsius
3 OXYGEN	54	Amount of oxygen during storage: 1 = 2% 2 = 6% 3 = 10%
4 ROT	54	Diameter of rotted area in mm

PRES

This file contains the number of years U.S. Presidents (since Abraham Lincoln) lived after first being inaugurated, and the life expectancy of a man the same age as each President was when he was inaugurated.

Column Name	Count	Description
1 EXPECTED	19	Life expectancy at first inauguration
2 ACTUAL	19	Actual years lived after first inauguration

PULSE

Students in an introductory statistics course participated in a simple experiment. Each student recorded his or her height, weight, gender, smoking preference, usual activity level, and resting pulse. Then they all flipped coins, and those whose coins came up heads ran in place for one minute. Then the entire class recorded their pulses once more.

Column Name	Count	Description
1 PULSE1	92	First pulse rate
2 PULSE2	92	Second pulse rate
3 RAN	92	1 = ran in place 2 = did not run in place
4 SMOKES	92	1 = smokes regularly 2 = does not smoke regularly
5 SEX	92	1 = male 2 = female
6 HEIGHT	92	Height in inches
7 WEIGHT	92	Weight in pounds
8 ACTIVITY	92	Usual level of physical activity: 1 = slight 2 = moderate 3 = a lot

RADON

Investigators measured radiation in an experimental chamber, using four different devices: filters, membranes, open cups, and badges. They tested 20 devices of each type, and recorded the amount of radiation that each device measured.

Column Name	Count	Description
1 FILTER	20	Radiation measured by filter devices
2 MEMBRANE	20	Radiation measured by membrane devices
3 OPEN CUP	20	Radiation measured by open cup devices
4 BADGE	20	Radiation measured by badge devices

REALEST

Investigators collected and summarized information on each home sold in Oxford, Ohio during 1987, including information on the asking and selling price and characteristics of the home.

Column Name	Count	Missing	Description
1 SELL $	51		Asking price of home
2 ASK $	51		Selling price of home
3 BATHS	51		Number of baths
4 BEDROOMS	51		Number of bedrooms
5 AGE	51	8	Age of home
6 BASEMENT	51	1	Basement? 1 = yes 2 = no
7 CARS	51		Size of garage (car capacity)

REPORTS

Many universities around the country are part of a small town or city. In many cases the student population dwarfs the town or city's full-time resident population. This file contains data on the total number of police reports for each month over the past six and a half years. A police report was defined as any arrest or any formal statement taken by the police department. Hence the reports cover the spectrum of charges and complaints; these include noise violations, burglary, and assaults.

Column Name	Count	Description
1 YEAR	80	Year
2 MONTH	80	Month
3 TOTAL	80	The total number of police reports for the month

RIVER1 - RIVER5

During the summer of 1988, one of the hottest on record in the midwest, a graduate student in Environmental Science did a study for a state Environmental Protection Agency. She studied the impact of an electric generating plant along a river by observing several water characteristics at five sites along the river. Site #1 was approximately four miles upriver from the electrical plant, about six miles downriver from a moderately large midwestern city, and directly downriver from a large suburb of this city. Site #2 was directly upriver from the cooling inlets of the plant. Site #3 was directly downriver from the cooling discharge outlets of the plant. Site #4 was approximately 3/4 of a mile downriver from site #3, and site #5 was approximately six miles downriver from site #3.

The EPA was most concerned with the plant's use of river water for cooling. They feared that the plant was raising the temperature of the water and hence endangering the aquatic species that lived in the river.

The student anchored data sounds (battery-operated and self-contained canisters that float in the river) at five sites and took hourly measurements of the temperature, pH, conductivity, and dissolved oxygen content. She left them there for five consecutive days.

The data sets RIVER1 through RIVER5 contain the recordings for each separate site.

Column Name	Count	Description
1 HOUR	*	The hour of the measurement in military time
2 TEMP	*	The temperature in °C of the river at the given hour
3 PH	*	The pH of the river at the given hour
4 COND	*	The conductivity in electrical potential of the river at the given hour
5 DO	*	The dissolved oxygen content of the river at the given hour

* The counts for the different data sets are: RIVER1=95; RIVER2=95; RIVER3=95; RIVER4=95; RIVER5=97

SALAMDER

Zoologists housed 14 salamanders in individual cages. During the period of study, they exposed the salamanders to normal conditions, monitoring each cage separately. They recorded the activity level in each cage hourly over the course of six days, for a total of 144 hourly assessments. These results were subsequently used to compare other experimental conditions (not included in this data set).

Column Name	Count	Description
1 ACTIVITY	144	The daily activity level for 14 salamanders

SALES

Consolidate Stores Inc. investigated the revenues from the women's sportswear division over the years 1988 through 1991. Seven days were randomly selected each year and the total revenue from this division was recorded.

Column Name	Count	Description
1 1991	7	The daily revenues of the women's sportswear division for 7 randomly selected days in 1991
2 1990	7	The daily revenues of the women's sportswear division for 7 randomly selected days in 1990
3 1989	7	The daily revenues of the women's sportswear division for 7 randomly selected days in 1989
4 1988	7	The daily revenues of the women's sportswear division for 7 randomly selected days in 1988

SCHOOLS

In December 1975, many public school teachers in Madison, Wisconsin called in sick as a protest against administrators.

Column Name	Count	Description
1 SCHLTYPE	48	School level: 1 = elementary 2 = middle 3 = high
2 TEACHERS	48	Number of teachers in the school
3 SICK	48	Number who called in sick

SNOWFALL

Using data collected at the weather station outside of a small midwest town, a college student recorded the yearly snowfall for the town for the past 15 years, back to 1976.

Column Name	Count	Description
1 SNOWFALL	15	The snowfall in inches for the past 15 winters in a small midwest town
2 RAIN	15	The equivalent rainfall in inches

SPEED2

One column of this data set contains speeds of cars while the other contains a speed control method identifier. See the CARSPEED data set for more information; it contains the speeds by method.

STA261

This is the complete grade information for 45 students in an introductory class in statistics, including the results of the 11 quizzes, the three hour exams, and the final exam. The quiz scores are all out of 15 possible and the four exams out of 100 possible points, with 10 extra credit points.

Column Name	Count	Description
1-11 Q1-Q11	45	Quiz scores
12-14 T1-T3	45	Hour exam scores
15 FINAL	45	Final exam scores

STA368

The single column of this data set contains the results of the first quiz (15 possible) in an introductory statistics course.

STEALS

Early in the 1991 major league baseball season, Rickey Henderson of the Oakland Athletics baseball team became the stolen base leader, passing Lou Brock. Newspapers reported the number of bases Mr. Henderson stole for each day of the week.

Column Name	Count	Description
1 DAY	7	The day of the week
2 BASES	7	The total number of bases that Mr. Henderson stole on the given day of the week

TBILLS

This file contains information on the bid and asking price and yield of treasury bills for 34 selected days in 1991.

Column Name	Count	Missing	Description
1 BID	34		The price bid for a treasury bill on a selected day
2 ASK	34		The asking price for a treasury bill on a selected day

3 CHANGE	34	5	The change in the price
4 YIELD	34		The yield of the bill on the selected day

TECHN

This data set contains the salary information on all salaried employees in four departments at the Technitron company. The data set PAY is a subset of this larger data set.

Column Name	Count	Description
1 SALARY	46	The salary of an employee
2 YRS EM	46	The number of years employed at Technitron
3 PRIOR YR	46	The number of prior years' experience
4 EDUC	46	Years of education after high school
5 ID	46	The company identification number for the employee
6 GENDER	46	The coded gender of the employee: 0 = female 1 = male
7 DEPT	46	The department of the employee: 1 = sales 2 = purchasing 3 = advertising 4 = engineering
8 SUPER	46	The number of employees supervised by this employee

TRANS

As part of a study investigating the quality of transistors produced at Flexible Circuits, Inc., 510 transistors were sampled from stock and were classified with respect to the day of the week the transistor was manufactured and the quality of the transistor. The data are stored in a contingency table format.

Column Name	Count	Description
1 (unnamed)	3	A description of the quality of the transistors
2 MON	3	The number of transistors manufactured on Monday, classified by the quality of the transistor
3 TUE	3	The number of transistors manufactured on Tuesday, classified by the quality of the transistor
4 WED	3	The number of transistors manufactured on Wednesday, classified by the quality of the transistor
5 THU	3	The number of transistors manufactured on Thursday, classified by the quality of the transistor
6 FRI	3	The number of transistors manufactured on Friday, classified by the quality of the transistor

TEST

This data set is the grade summary for a social studies class in a secondary school. Three test scores have been recorded for the 24 students in the class.

Column Name	Count	Description
1 LAST NAM	24	The last name of the student
2 FIRST	24	The first name of the student

3 TEST1	24	The score (out of 100) on the first test
4 TEST2	24	The score (out of 100) on the second test
5 TEST3	24	The score (out of 100) on the third test

TREES

The diameter, height, and volume of 31 black cherry trees in Allegheny National Forest are recorded in this file.

Column Name	Count	Description
1 DIAMETER	31	Diameter of tree in inches (measured at 4.5 feet above ground)
2 HEIGHT	31	Height of tree in feet
3 VOLUME	31	Volume of tree in cubic feet

TVVIEW

Ten married couples were randomly selected and the husband and wife were asked the average amount of time (in hours) each spent watching TV per week.

Column Name	Count	Description
1 HUSBAND	10	The average number of hours spent watching TV per week by the husband
2 WIFE	10	The average number of hours spent watching TV per week by the wife
3 TOTAL	10	The average number of hours spent watching TV per week by the husband and wife
4 DIFF	10	The difference between the average number of hours spent watching TV per week by the husband and wife

TWAIN

Analysis of writing style can help determine the author of anonymous writing, by comparing certain qualities of the text to those of known works. One aspect that can be numerically analyzed is the frequency of words of certain lengths found in the writing. Claude S. Brinegar performed such a study on ten letters, thought to be written by Mark Twain, which were published in 1861 in a New Orleans newspaper, signed "Quintus Curtius Snodgrass." In the data file, the first three letters are grouped in the first column, the second three in the second column, and the last four in the third column; the remaining columns contain data of known Twain writings. In any given column, the first entry is the number of two-letter words in the text, the second is the number of three-letter words, and so on, up to the twelfth entry, the number of words of 13 or more letters.

Column Name	Count	Description
1 S.1-3	12	Data from the first three letters of Snodgrass
2 S.4-6	12	Data from the second three letters of Snodgrass
3 S.7-10	12	Data from the final four letters of Snodgrass
4 T.1861	12	Data from two of Twain's letters, 1858 and 1861
5 T.1863	12	Data from four of Twain's letters, 1863
6 T.1867	12	Data from one of Twain's letters, 1867
7 T.1872	12	Data from Twain's *Roughing It*, 1872
8 T.1897	12	Data from Twain's *Following the Equator*, 1897

USDEMOG

Census values for the U.S. are given, including the estimated population and population density.

Column Name	Count	Missing	Description
1 YEAR	35		Year
2 US POPLN	35	14	Population of the U.S. in millions
3 POP DENS	35	14	Population density of the U.S. in people per square mile

UTILITY

A family monitored its utility usage for approximately four years, recording electricity and water usage, as well as cost, on a monthly basis.

Column Name	Count	Missing	Description
1 MONTH	44		Month
2 YEAR	44		Year
3 KWH	44		Kilowatts of electricity used
4 KWH $	44		Cost of electricity
5 GAL	44	2	Gallons of water used
6 GAL $	44	2	Cost of water

VOICE

A professor in a theater department was interested in determining if a particular voice training class improved a performer's voice quality. The professor studied 20 students, 10 of whom took the class and 10 of whom did not. Six judges rated the subjects' voice quality on a scale of 1-6 (6 = best) before and after the class. The data set includes scores for the ten subjects who took the class.

Column	Name	Count	Description
1	1 BEFORE	6	Subject 1's score before the class (1-6, 6 is best)
2	1 AFTER	6	Subject 1's score after the class
3	2 BEFORE	6	Subject 2's score before the class
4	2 AFTER	6	Subject 2's score after the class
etc. for columns 5-18			etc. for columns 5-18
19	10 BEFOR	6	Subject 10's score before the class
20	10 AFTER	6	Subject 10's score after the class

VPVH

Each year the Nielsen company does an extensive survey of TV viewing habits of the U.S. population. Many of the surveys are done by age and gender groups, because sponsors are very interested in how these variables affect what programs people watch.

This data set contains the VPVH's estimates for several different categories of television programming. The VPVH estimate is the estimated number of viewers per 1000 viewing households tuned to a station or program. This data set includes the ten VPVH estimates for each of four commonly surveyed gender/age groups (men age 18-34, women age 18-34, men age 55+, and women 55+) for the following kinds of programs: network movies, situation comedy, and sports. For the sports program, only data for the two male age groups were given.

Column Name	Count	Missing	Description
1 MOVIES	40		The VPVH data for top ten network movies
2 COMEDY	40		The VPVH data for top ten situation comedies
3 SPORTS	40	20	The VPVH data for top ten sporting events
4 GROUP	40		The gender/age group
5 GROUP N	40		The gender age/group coded numerically

WGTHGT

A young college graduate in his first teaching assignment in a local school district decided to introduce his first-grade classes to statistics. As a class exercise, he brought in a scale and a tape measure and recorded the heights (in inches) and weights (in pounds) of all 73 students in his two homeroom classes.

Column Name	Count	Description
1 WEIGHT	73	Weight in pounds of first grader
2 HEIGHT	73	Height in inches of first grader

Appendix C: Quick Reference

Contains list of all STAT101 commands & their syntax, plus keyboard shortcuts and functions. Commands are left justified; subcommands are indented.

NOTATION

K denotes a constant, such as **8.3** or **K14**

C denotes a column, such as **C12** or **'Height'**

E denotes either a constant or column

[ ] encloses an optional argument

1 SYMBOLS

* *Missing Value Symbol.* An * can be used as data in READ, SET, and INSERT; in data files; and in the Data Editor. Enclose the * in single quotes in session commands and subcommands.

*Comment Symbol.* The symbol # anywhere on a line tells STAT101 to ignore the rest of the line.

& *Continuation Symbol.* To continue a command onto another line, end the first line with the symbol &. You can use ++ as a synonym for &.

2 WORKSHEET AND COMMANDS

STAT101 consists of a worksheet for data and commands that allow you to enter data into the worksheet; manipulate and transform your data; produce graphical and numerical summaries; and perform a wide range of statistical analyses.

The worksheet contains columns of data, denoted by column numbers **C1, C2, C3...**, or by column names such as **'Age'** or **'Height'**. Enclose column names in single quotation marks in session commands. The worksheet can also contain stored constants, denoted by **K1, K2, K3**.

To execute a command, choose it from a menu or enter it into the Session screen at the STAT101 prompt, STAT>. When you execute a menu command, STAT101 lists available options in dialog boxes. When you execute a session command, you need to know the correct command syntax.

3 SUBCOMMANDS

Some STAT101 session commands have subcommands. To use a subcommand, put a semicolon at the end of the main command line. Then type the subcommands, starting each on a new line and ending each with a semicolon. When you are done, end the last subcommand with a period.

The subcommand **ABORT** cancels the whole command.

4 KEYBOARD SHORTCUTS

[Esc] or [Alt] + [M] to switch to Session screen

[Esc] or [Alt] + [D] to switch to Data Editor

[F1] for HELP in Data Editor

[Ctrl] + [Break] to interrupt a macro

From Data Editor

[Ctrl] + [Home] to move to beginning of worksheet

[Ctrl] + [End] to move to end of worksheet

[Ctrl] + [→] to move one screen to the right

[Ctrl] + [←] to move one screen to the left

[F2] to edit a value within a cell

[F3] to change entry direction

[F5] to go to a specific row and/or column

[F6] to move to beginning of next row or column, depending on data entry direction

5 GENERAL INFORMATION

HELP explains STAT101 commands, can be a command or a subcommand

INFO [C...C] gives status of worksheet

STOP ends current session

6 INPUT AND OUTPUT OF DATA

READ data [from 'filename'] into C...C
SET data [from 'filename'] into C
INSERT data [from 'filename'][between rows K and K] of C...C

READ, SET, and **INSERT** have the subcommands
   FORMAT (Fortran format)
   NOBS =K

END of data (optional)
NAME for C is 'name', for C is 'name'... for C is 'name'
PRINT the data in E...E
WRITE [to 'filename'] the data in C...C

PRINT and **WRITE** have the subcommand
   FORMAT (Fortran format)
SAVE [in 'filename'] a copy of the worksheet
   PORTABLE
RETRIEVE the STAT101 saved worksheet [in 'filename']
   PORTABLE

7 EDITING AND MANIPULATING DATA

LET C(K) = K # changes the number in row K of C
DELETE rows K...K of C...C
ERASE all data in E...E
INSERT (see Section 6)
COPY C...C into C...C
COPY C into K...K
   USE rows K...K
   USE rows where C = K...K
   OMIT rows K...K
   OMIT rows where C = K...K
COPY K...K into C
CODE (K...K) to K ... (K...K) to K for C...C, put in C...C
STACK (E...E) ... on (E...E), put in (C...C)

SUBSCRIPTS *into* C
UNSTACK (C...C) *into* (E...E) ... (E...E)
    SUBSCRIPTS *are in* C
CONVERT *using table in* C C, *the data in* C, *and put in* C
CONCATENATE C...C *put in* C

---

8 ARITHMETIC

LET = *expression*

Expressions may use:
Arithmetic operators   + - * / ** (exponentiation)
Comparison operators   = ~= < > <= >=
Logical operators   & | ~
and any of the following: ABSOLUTE, SQRT, LOGTEN, LOGE, EXPO, ANTILOG, ROUND, SIN, COS, TAN, ASIN, ACOS, ATAN, SIGNS, NSCORE, PARSUMS, PARPRODUCTS, COUNT, N, NMISS, SUM, MEAN, STDEV, MEDIAN, MIN, MAX, SSQ, SORT, RANK, LAG, EQ, NE, LT, GT, LE, GE, AND, OR, NOT.

Simple Arithmetic Operations

ADD E *to* E ... *to* E, *put in* E
SUBTRACT E *from* E, *put in* E
MULTIPLY E *by* E ... *by* E, *put in* E
DIVIDE E *by* E, *put in* E
RAISE E *to the power* E, *put in* E

Columnwise Functions

ABSOLUTE *value of* E, *put in* E
SQRT *of* E, *put in* E
LOGE *of* E, *put in* E
LOGTEN *of* E, *put in* E
EXPONENTIATE E, *put in* E
ANTILOG *of* E, *put in* E
ROUND *to integer* E, *put in* E
SIN *of* E, *put in* E
COS *of* E, *put in* E
TAN *of* E, *put in* E
ASIN *of* E, *put in* E
ACOS *of* E, *put in* E
ATAN *of* E, *put in* E
SIGNS *of* E, *put in* E

PARSUMS *of C, put in C*
PARPRODUCTS *of C, put in C*

Normal Scores

NSCORES *of C, put in C*

Columnwise Statistics

COUNT *the number of values in* **C** *[put in* **K***]*
N *(number of nonmissing values) in* **C** *[put in* **K***]*
NMISS *(number of missing values) in* **C** *[put in* **K***]*
SUM *of the values in* **C** *[put in* **K***]*
MEAN *of the values in* **C** *[put in* **K***]*
STDEV *of the values in* **C** *[put in* **K***]*
MEDIAN *of the values in* **C** *[put in* **K***]*
MINIMUM *of the values in* **C** *[put in* **K***]*
MAXIMUM *of the values in* **C** *[put in* **K***]*
SSQ *(uncorrected sum of sq.) for* **C** *[put in* **K***]*

Rowwise Statistics

RCOUNT *of* **E**...**E** *put in* **C**
RN *of* **E**...**E** *put in* **C**
RNMISS *of* **E**...**E** *put in* **C**
RSUM *of* **E**...**E** *put in* **C**
RMEAN *of* **E**...**E** *put in* **C**
RSTDEV *of* **E**...**E** *put in* **C**
RMEDIAN *of* **E**...**E** *put in* **C**
RMINIMUM *of* **E**...**E** *put in* **C**
RMAXIMUM *of* **E**...**E** *put in* **C**
RSSQ *of* **E**...**E** *put in* **C**

9 PLOTTING DATA

STEM-AND-LEAF *display of* **C**...**C**
  TRIM *outliers*
  INCREMENT = **K**
  BY **C**

BOXPLOT *for* **C**
  INCREMENT = **K**
  START *at* **K** *[end at* **K***]*
  BY C

        LINES = K
        NOTCH [K% confidence] sign c.i.
        LEVELS K...K

HISTOGRAM C...C

DOTPLOT C...C

    HISTOGRAM and **DOTPLOT** have the subcommands
        INCREMENT = K
        START *at* K [*end at* K]
        BY C
        SAME *scales for all columns*

PLOT C *vs* C

MPLOT C *vs* C, *and* C *vs* C, *and* ... C *vs* C

LPLOT C *vs* C *using tags in* C

TPLOT C *vs* C *vs* C

PLOT, MPLOT, LPLOT and **TPLOT** have the subcommands
    YINCREMENT = K
    YSTART *at* K [*end at* K]
    XINCREMENT = K
    XSTART *at* K [*end at* K]

TSPLOT [*period* K] *of* C
    ORIGIN = K
    INCREMENT = K
    START *at* K [*end at* K]
    TSTART *at* K [*end at* K]

WIDTH *of all plots that follow is* **K** *spaces*
HEIGHT *of all plots that follow is* **K** *lines*

---

10 BASIC STATISTICS

DESCRIBE C...C
    BY C

ZINTERVAL [K% *confidence*] *assuming sigma* = K *for* C...C
ZTEST [*of mu* = K] *assuming sigma* = K *for* C...C
    ALTERNATIVE = K

TINTERVAL [K% *confidence*] *for data in* C...C
TTEST [*of mu* = K] *on data in* C...C
    ALTERNATIVE = K

TWOSAMPLE *test and c.i. [K% confidence] samples in* **C C**
    ALTERNATIVE = K
    POOLED *procedure*

TWOT *test and c.i. [K% confidence] data in* **C**, *groups in* **C**
    ALTERNATIVE = K
    POOLED *procedure*

CORRELATION *between* **C...C** *[put in* **M**]

---

11 REGRESSION

REGRESS C *on* **K** *predictors* **C...C** *[put st. resids in* **C** *[fits in* **C**]]
    NOCONSTANT *in equation*
    COEFFICIENTS *put in* **C**
    RESIDUALS *put in* **C** *(observed - fit)*
    PREDICT *for* **E...E**
    DW *(Durbin-Watson statistic)*

STEPWISE *regression of* **C** *on the predictors* **C...C**
    FENTER = K *(default is four)*
    FREMOVE = K *(default is four)*
    FORCE C...C
    ENTER C...C
    REMOVE   C...C
    BEST K *alternative predictors (default is zero)*
    STEPS = K *(default depends on output width)*

NOCONSTANT *in* **REGRESS** *and* **STEPWISE** *commands that follow*
CONSTANT *fit a constant in* **REGRESS** *and* **STEPWISE**
BRIEF K

---

12 ANALYSIS OF VARIANCE

AOVONEWAY *aov for samples in* **C...C**
ONEWAY *aov, data in* **C**, *levels in* **C** *[put resids in* **C** *[fits in* **C**]]

TWOWAY *aov, data in* **C**, *levels in* **C C** *[put resids in* **C** *[fits in* **C**]]
    ADDITIVE *model*
    MEANS *for the factors* **C** [**C**]

---

13 NONPARAMETRICS

RUNS *test [above and below* **K**] *for* **C**
STEST *sign test [median =* **K**] *for* **C...C**

ALTERNATIVE = K
SINTERVAL *sign c.i. [K% confidence] for* **C...C**
WTEST *Wilcoxon one-sample rank test [median = **K**] for* **C...C**

   ALTERNATIVE= K
WINTERVAL *Wilcoxon c.i. [K% confidence] for* **C...C**
MANN-WHITNEY *test and c.i. [K% confidence] on* **C C**
KRUSKAL-WALLIS *test for data in* **C**, *subscripts in* **C**
WALSH *averages for* **C**, *put in* **C** *[indices into* **C C**]
WDIFF *for* **C** *and* **C**, *put in* **C** *[indices into* **C C**]
WSLOPE *y in* **C**, *x in* **C**, *put in* **C** *[indices into* **C C**]

---

14 TABLES

TALLY *the data in* **C...C**
  COUNTS
  PERCENTS
  CUMCOUNTS *cumulative counts*
  CUMPERCENTS *cumulative percents*
  ALL *four statistics above*

CHISQUARE *test on table stored in* **C...C**
TABLE *the data classified by* **C...C**
  MEANS *for* **C...C**
  MEDIANS *for* **C...C**
  SUMS *for* **C...C**
  MINIMUMS *for* **C...C**
  MAXIMUMS *for* **C...C**
  STDEV *for* **C...C**
  STATS *for* **C...C**
  DATA *for* **C...C**
  N *for* **C...C**
  NMISS *for* **C...C**
  PROPORTION *of cases = **K** [through **K**] in* **C...C**
  COUNTS
  ROWPERCENTS
  COLPERCENTS
  TOTPERCENTS
  CHISQUARE *analysis [output code = **K**]*
  MISSING *level for classification variable* **C...C**
  NOALL *in margins*
  ALL *for* **C...C**
  FREQUENCIES *are in* **C**

15 TIME SERIES

ACF [*with up to* K *lags*] *for series in* C [*put in* C]
PACF [*with up to* K *lags*] *for series in* C [*put in* C]
CCF [*with up to* K *lags*] *between series in* C *and* C
DIFFERENCES [*of lag* K] *for data in* C, *put in* C
LAG [*by* K] *for data in* C, *put in* C
ARIMA $p = K, d = K, q = K$, *data in* C [*put resids in* C [*preds in* C [*coefs in* C]]]
ARIMA $p = K, d = K, q = K, P = K, D = K, Q = K, S = K$, *data in* C [*put resids in* C [*preds in* C [*coefs in* C]]]
  CONSTANT *term in model*
  NOCONSTANT *term in model*
  STARTING *values are in* C
  FORECAST [*origin* = K] *up to* K *leads* [*put in* C [*limits in* C C]]

16 DISTRIBUTIONS AND RANDOM DATA

RANDOM K *observations into* C...C
  BERNOULLI *trials* $p = K$
PDF *for values in* E [*put results in* E]
CDF *for values in* E [*put results in* E]
INVCDF *for values in* E [*put results in* E]

RANDOM, PDF, CDF, INVCDF have the subcommands
  BINOMIAL $n = K, p = K$
  POISSON $mu = K$
  INTEGER *discrete uniform on integers* K *to* K
  DISCRETE *dist. with values in* C *and probabilities in* C
  NORMAL [$mu = K$ [$sigma = K$]]
  UNIFORM [*continuous on the interval* K *to* K]
  T *degrees of freedom* = K
  F *df numerator* = K, *df denominator* = K

SAMPLE K *rows from* C...C *put in* C...C
BASE *for random number generator* = K

17 SORTING

SORT C [*carry along* C...C] *put in* C [*and* C...C]

RANK *the values in* C, *put ranks into* C

18 MISCELLANEOUS

OUTFILE 'filename' *put all output in file*
    OW = K *output width of file*
    OH = K *output height of file - PC only*
    NOTERM *no output to terminal*
NOOUTFILE *output to terminal only*

NOTE *comments may be put here*

OW = K *number of spaces for width of output*
IW = K *number of spaces for width of input*
BRIEF = K *controls amount of output from* **REGRESS** *and* **ARIMA**

RESTART *begin fresh STAT101 session*

PAPER *output to printer*
    OW = K *output width of printer*
    OH = K *output height of printer*
    NOTERM *no output to terminal*
NOPAPER *output to terminal only*
NEWPAGE *start next output on a new page*
SYSTEM *provides access to operating system commands*
CD [*path*] *change directory*
DIR [*path*] *list files*
TYPE [*path* [**filename**]] *display contents of a file*

---

19 STORED COMMANDS AND LOOPS

The commands **STORE** and **EXECUTE** provide the capability for simple macros (stored command files) and loops.

EXECUTE 'filename' [K *times*]
STORE [in **'filename'**] *the following commands (enter commands)*
END *of storing commands*
NOECHO *the commands that follow*
ECHO *the commands that follow*

The CK capability. The integer part of a column number may be replaced by a stored constant.

    EXAMPLE: LET K1 = 5
                      PRINT C1 - CK1
since K1 = 5, this PRINTS C1 through C5.

Appendix D: MINITAB Software Products

Documentation

STAT101 users who need more statistical power and space for more data will naturally turn to MINITAB statistical software, which is fully compatible with STAT101 file formats, command structure, and style. Minitab Inc. offers a variety of clearly written documentation to assist new MINITAB users in installing, running, and applying MINITAB functionality to statistical problem solving.

MINITAB Reference Manual

- Release 9 for Windows
- Release 9
- Release 8, PC Version
- Release 8, Macintosh Version
- Release 7

A complete and detailed manual covering all commands, step-by-step instructions, sample sessions, and system-specific information. The *MINITAB Reference Manual* for Release 9 documents all of the features available in the Standard Version of MINITAB and MINITAB for Windows.

MINITAB Graphics Manual

- Release 9 for Windows
- Release 9

A comprehensive guide, filled with examples and detailed information on how to use Professional Graphics commands, available with the Enhanced Version of MINITAB and MINITAB for Windows.

MINITAB QC Manual

- Release 9 for Windows
- Release 9

A comprehensive guide on how to use MINITAB's quality control and improvement features, including statistical process control charts and design of experiments, capabilities, available with the Enhanced Version of MINITAB and MINITAB for Windows.

MINITAB QC Supplement, Release 8

A supplement to the Release 8 *MINITAB Reference Manual* that documents additional quality control and improvement capabilities included in Release 8 of MINITAB. (Automatically included with purchase of Release 8 *MINITAB Reference Manual*: not available independently.)

MINITAB Quick Reference Card

- Release 9 for Windows
- Release 9
- Release 8, PC Version
- Release 8, Macintosh Version
- Release 7

A convenient small card that summarizes all MINITAB commands.

MINITAB Handbook, Second Edition

A supplementary text that teaches basic statistics using MINITAB. The Handbook features the creative use of plots, application of standard statistical methods to real data, in-depth exploration of data, and more. (For more than 5 copies of this text, contact Wadsworth Publishing: in the USA, call 800-343-2204. Outside the USA, call 415-595-2350; ask for International Department.)

Companion Textbook List

More than 250 textbooks and textbook supplements that include MINITAB have been published. For a complete bibliography, organized by discipline and level (introductory, intermediate, and advanced), contact Minitab Inc.

To order, contact Minitab Inc., 3081 Enterprise Dr., State College, PA 16801-3008, U.S.A.

Software

MINITAB Statistical Software

- Release 9 for Windows
- Release 9

- Release 8 for DOS
- Release 8 for Macintosh

One of the leading statistics packages worldwide, with MINITAB Statistical Software you can analyze your data with power, flexibility, and convenience. For example, you can

- present your data with colorful, high-resolution graphs
- analyze large amounts of data
- use menus and dialog boxes rather than type your commands
- do more advanced and complete analysis
- transfer data easily to and from other applications using Copy and Paste commands or a custom Lotus 1-2-3 function

To order, contact Mintab Inc., 3081 Enterprise Dr., State College, PA 16801-3008, U.S.A.

Student Edition of MINITAB

- Release 8 for DOS
- Release 8 for Macintosh

A streamlined and lower-cost version of the commercial program with a manual written specifically for students, including case studies and 14 hands-on, self-paced tutorials using a large collection of data sets drawn from business, social sciences, life sciences, and engineering.

To order, contact Addison-Wesley Publishing Company, Inc., 1 Jacob Way, Reading, MA 01867.

Index

\# comment, 16, 59, 83
& continuation, 58, 83
* missing value, 21, 58, 65, 82
+ key combination, 2, 19
1STGRADE.MTW, 225
\ disk directory level, 62
\ problem accessing Data Editor, 18
π pi stored constant, 49

Abbreviate
   Command name, 20
   Range, 174
ABORT, 56
Abort error message, 221
ABSOLUTE, 84
ACCOUNTS.MTW, 225
ACF autocorrelation, 84
ACID.MTW, 226
ACOS arccosine, 85
Active cell, 19, 41
ADD columns or constants, 85
Advanced sample session, 28
Add data to existing column, 125
AGGRESS.MTW, 227
ALFALFA.MTW, 227
Alpha data, 50
   Combining two columns into one, 107
   Converting to numeric, 108
   Format to export or import, 65
   Missing values, 82
   Reformatting, 50
Analysis of variance
   Commands, 77
   Sample session, 30
Annotating your work, 16, 83, 144
ANTILOG, 86
AOVONEWAY one-way analysis of variance, 86, 145
Arguments, 53, 56, 82
ARIMA, 87, 101
Arithmetic functions, 91, 93, 95

Arrow keys, turning off Num Lock key, 19
ASCII file, 64
   Exporting, 65
   Importing, 64
   Types used by STAT101, 70
   Viewing contents, 208
ASIN arcsine, 97
ATAN arctangent, 97
Autocorrelation, 84
AUTOEXEC.BAT file, 6, 223
Autoregressive integrated moving average, see ARIMA
AZALEA.MTW, 228

Backing up STAT101 program disk, 9
BANKING.MTW, 228
BASE for random number generation, 98
Basic descriptive statistics, 27
   Commands, 76
Basic sample session, 14
Binary worksheet file, 70
Boolean operations, 93
BOXPLOT, 98
   Sample session, 30, 32
BPRES.MTW, 164, 229
Brackets, 57, 82
BRIEF output display, 101

C for Column, 56, 82
CAMSHAFT.MTW, 229
Can't Find File error message, 222
Can't Start STAT101 error message, 223
Cancel command, 56
Capital letters, 57, 82
CARSPEED.MTW, 87, 204, 229
CARTOON.MTW, 230
Case of letters in commands, 2
   see also Capital letters, 82
Categorical data, 189

CCF cross correlation, 102
CD change directory (DOS command), 10, 103
CD1.MTW, 231
CD2.MTW, 118, 232
CDF cumulative distribution function, 103
Cell, 19, 41
Changing screen colors, 6
Chisquare statistic, 194
Chisquare test, 189
CHISQUARE test of association, 104
CHOLEST.MTW, 232
CITIES.MTW, 233
Clearing the worksheet, 14 see also RESTART
CLOTTING.MTW, 234
CODE, 106
  Sample session, 21
Color monitor, 6
Column, 39
  Adding data to existing column, 125
  Arithmetic and Transformations, 91
  Changing data format, 50
  Code, 106
  Combining two alpha columns into one, 107
  Copying, 109
  Creating from existing column, 31
  Functions, 95
  Naming, 47
  Percent, 193
  Sorting, 177
  Subsets, 31
Combine data from separate files, 65
Command
  Cancel, 56
  Equivalent methods of issuing, 57
  Error while typing, 13
  Issuing, 53
  Operating on missing values, 83
  Quick reference, 273
  Reference, 71
  Saving in outfile, 67
  Shortcut, 55
  Subcommands, 55
  Syntax, 54, 57, 82
  Typographical conventions, 56
Comment, adding to output, see # comment
Comparing variable levels, 35
Comparison operations, 92
COMPUTER.MTW, 234
CONCATENATE alpha columns, 107
Confidence interval
  Analysis of variance, 29, 208
  ARIMA, 87
  Boxplot, 98
  Mann-Whitney, 136
  Regression, 165
  SINTERVAL, 176
  T-test, 202
  TINTERVAL, 198
  Used in simulations, 187
  WDIFF, 212
  WINTERVAL, 213
  ZINTERVAL, 217
CONFIG.SYS file, 6
CONSTANT term, 107
Constants, 39, 49
CONT> prompt, 59
Continue? prompt, 59, 145
Continuing a command line, see & continuation
CONVERT alpha and numeric data, 108
COPY, 109
  Sample session, 31
Copying commands, 25
Copying STAT101 program disk, 9
COREDATA.MTW, 235
Correcting data errors, 23
CORRELATE, 111
COSINE, 112
COUNT number of values in a column, 112
CRANKSH.MTW, 235
Cross correlation, 102
Cross tabulation, 189
CROWDS.MTW, 235

Cumulative distribution function, 103
Cursor, 46

DAT file, see also ASCII file, 65, 70, 215
Data
  Code, 106
  Combining from separate files, 65
  Commands, 74
  Converting alpha and numeric, 108
  Copying, 109
  Exporting to other applications, 65
  File, see ASCII file
  Formats, 49
  Importing from other applications, 64
  Inserting row, 125
  Missing values, see * missing value
  Paired, 203
  Plotting commands, 75
  Saving in Sample session, 24
  Sorting, 177
  Subset using logical operators, 93
  Tables, 189
  Viewing, 72, 155
Data Editor, 39
  Accessing, 39
  Adding data, 23
  Changing screen colors, 7
  Data-entry arrow, 25
  Edit mode, 46
  Editing the worksheet, 46
  Entering data, 44
  Entry mode, 46
  Help, 43
  Keyboard summary, 51
  Menu, 42
  Moving around, 41
  Problem accessing, 18
  Sample session, 20
Data entry
  Adding data to existing column, 23, 125
  By column or by row, 23
  Commands, 72

Correcting mistakes, 24
  Data Editor, 44, 45
  From the keyboard, 125
  Patterned, 174
Data sets, see Sample data sets
Data-entry arrow, 23, 44
DATA> prompt, 59
Default, 5
DEL (DOS command), 67
DELETE rows from a worksheet, 113
Deleting files, 67
DESCRIBE, 114
  Sample Session, 27
Descriptive statistics, 27
  Commands, 76
DIFFERENCES, 116
DIR (DOS command), 116
Directory, 3
  Changing, 10
  Listing all files, 116
Discrete distributions, 104
Disk
  Copying program disk, 9
  Drives, 3
  High density, 1
  Saving worksheet to a floppy, 63
  Space for STAT101, 1, 6
  Write-protection, 4
Distributions
  Commands, 80
  Cumulative probabilities, 103
  Inverse cumulative distribution function, 127
  Probability density function, 151
  RANDOM, 157
DIVIDE, 117
DOS
  Access from STAT101 using SYSTEM, 188
  CD command (Change Directory), 10, 103
  Commands, 73
  DEL command, 67
  DIR command (Directory), 116
  MORE command, 221
  Prompt, 3, 10
  Release requirements, 1

TYPE command, 147, 208
  Using, 3
DOTPLOT, 117
  Sample session, 19

E for Argument, 57, 82
e stored constant, 49
ECHO, 119
Edit mode, 46
Editing data, see also Data Editor
  Commands, 74
EMPLOY.MTW, 236
END, 120
Enter key, 2
Entering data, see Data entry
Entry mode, 46
ERASE, 120
Error
  Messages, 221
  While entering data, 24
  While typing a command, 13
Escape key producing \, 18
EXAM.MTW, 236
EXAMPLE.MTW, 108, 110, 126,
  172, 178, 179, 198, 210, 211, 216,
  236
EXECUTE commands in a
  macro, 70, 121
Exiting STAT101, 11, 187
Exponential data format, 50
EXPONENTIATE, 121
Exporting data, 65

FA.MTW, 237
FABRIC.MTW, 237
Fail error message, 221
FALCON.MTW, 238
File extension, 61, 63
File, see also ACII file
  Combining, 65
  Commands, 73
  Deleting, 67
  Handling, 61
  Name problem, 222
  Naming, 62
  Types 61, 70
FISH.MTW, 238
FLEX2.MTW, 240
FLEXIBLE.MTW, 239

Floating point data format, 50
FOOTBALL.MTW, 241
Forecasts for a time series, 88
Formatting data, 49
Functions
  Arithmetic, 93
  Column, 95
  Row, 96
FURNACE.MTW, 241

GENETICS.MTW, 243
Getting help, see Help
Global changes using CODE, 21
GOLF.MTW, 243
GoTo, 24
  Sample session, 24
GRADES.MTW, 244
Graphing data, see Plotting data
GUARD.MTW, 244

HCC.MTW, 245
HEALTH.MTW, 245
HEART.MTW, 246
HEIGHT of plots, 122
HEIGHTS.MTW, 246
HELP, 58, 122
  Sample session, 25
Help
  Data Editor, 43
  Session screen, 26, 58
Highlighted cell, 19
HISTOGRAM, 55, 123
HOMESALE.MTW, 246
HWASTE.MTW, 247

Importing data, 64
INFO, 21, 125
  On selected columns, 21
  Sample session 17, 32
INSERT, 70, 125
Installing STAT101, 4
Integer data format, 50
INVCDF inverse cumulative
  distribution function, 127, 195
Issuing commands, 53
IW input width, 129

K for Constant, 49, 56, 82
Key combination, 2, 19
Keyboard summary for Data

Editor, 51
KRUNCHY.MTW, 247
KRUSKAL-WALLIS, 130

Labeled plot, see LPLOT
LAG, 131
LAKE.MTW, 248
LET, 91, 132
    Sample session, 31
License agreement, 1
LIS file, 67, 70
LOGE logarithm base e, 132
Logical operations, 92
LOGTEN logarithm base 10, 133
LOTTO.MTW, 248
LPLOT labeled plot, 133
    Sample session, 37

Macros, 121, 187
    Commands, 81
Manipulating data
    Commands, 74
MANN-WHITNEY, 136
MAPLE.MTW, 249
Math coprocessor, 1
Mathematical functions, 91
MAXIMUM, 137
Maximum value in column using DESCRIBE, 115
MEAN, 138
Mean of column using DESCRIBE, 114
MEATLOAF.MTW, 249
MEDIAN, 138
Median of column using DESCRIBE, 114
MIAMI.MTW, 115, 191, 196, 205, 208, 250
MINIMUM, 138
Minimum value in column using DESCRIBE, 115
MINITAB software, v, 283
Missing values, see * missing value
Mistake while typing a comand, 13
Monitor
    Changing screen colors, 6
    Requirements, 1

Monochrome monitor, 41
    Appearance of active cell, 41
Monochrome monitor, 6
MORE (DOS command), 221
More? prompt, 59
Mouse not supported, 2
MPLOT plot multiple pairs of columns, 138
    Sample session, 36
MSETUP, 6
MTB file, 70, 187
MTP file 70
MTW file, 62, 63, 70
Multiple scatter plots, see MPLOT
MULTIPLY, 140

N number of nonmissing values, 140
NAME, 140
    Sample session, 34
Naming columns, 47 see also NAME
Network, STAT101 not running on one, 8
NEWPAGE, 141
NEWPULSE.MTW, 24, 29
NIELSEN.MTW, 252
NMISS number of missing values, 142
NOCONSTANT, 142
NOECHO, 142
Nonparametric statistics
    Commands, 78
NOOUTFILE, 37, 67, 70, 143
    Sample session, 27 ,37
NOPAPER, 68, 143
Not Reading Drive error message, 221
NOTE, 144
NSCORE normal scores, 144
Numeric data, 50
    Convert to alpha data, 108

OH output height, 59, 145
ONEWAY analysis of
    Sample session, 29
OUTFILE, 67, 69, 70, 141, 146
    Sample session, 17

Outfile
   Annotating, 17, 83
   Printing 69
   Saving commands, 67
   Saving session, 67
   Viewing, 147
OW output width, 148

PACF partial autocorrelation, 148
Paired data
   T-test, 203
Pairwise deletion, 112
PAPER, 68, 141, 149
PAPER.MTW, 253
PARPRODUCTS partial
   products, 151
PARSUMS partial sums, 151
Partial autocorrelation, 148
PASTA.MTW, 254
Path name, 62
Patterned data entry, 174
PAY.MTW, 254
PDF probability distribution
   function, 151
Pearson product moment, 111
Personalizing STAT101, 4
PERU.MTW, 255
PLOT, 53
   Sample session, 37
Plotting data
   Commands, 75
PLYWOOD.MTW, 256
POPLAR1.MTW, 256
POPLAR2.MTW, 257
Portable worksheet files, 70
POTATO.MTW, 225, 257
PRES.MTW, 258
PRINT, 69, 155
Printing, 68
   Commands, 73
   Session, 149
   Worksheet in a compact form, 70
Probability density function, 151
Program, see Macros
Prompt, 59
   CONT>, 59
   Continue?, 59, 145
   DATA>, 59
   More?, 59

   STAT>, 11, 16, 20, 59
   SUBC>, 20, 55, 56
PULSE.MTW, 114, 193, 194, 258
   Used in sample sessions, 15
Punctuation for command
   language, 56

Quartile, 115
Quick reference to commands, 273

RADON.MTW, 259
RAISE value to a power, 156
RAM requirements, 1
RANDOM, 98, 157
Random data
   Commands, 80
   Distributions, 157
RANK, 112, 160
Rank correlation, 112
RCOUNT number of values in
   each row, 161
READ, 64, 70, 161
   Data format, 51
READ.ME text file for
   troubleshooting, 221
REALEST.MTW, 99, 260
REGRESS, 101, 107, 142, 163
   Sample session, 35
Regression, 35
   Commands, 77
REPORTS.MTW, 260
RESTART, 14, 40, 65, 166
RETRIEVE, 54, 63, 70, 166
   Sample session, 17
Retry error message, 222
Return key, 2
RIVER1.MTW-RIVER5.MTW,
   102, 261
RMAXIMUM maximum value in
   each row, 167
RMEAN mean of each row, 168
RMEDIAN median of each row,
   168
RMINIMUM minimum of each
   row, 168
RN number of nonmissing values
   in each row, 168
RNMISS number of missing
   values in each row, 169

ROUND, 169
Row, 39
Row functions, see R + individual command
RSSQ uncorrected sum or squares, 169
RSTDEV row standard deviation, 170
RSUM row sum, 170
RUNS test, 170

SALAMDER.MTW, 262
SALES.MTW, 262
Sample data sets
   Descriptions, 225
   Retrieving, 64
SAMPLE randomly from a column, 171
Sample session, 13
   Advanced, 28
   Basic, 14
SAVE, 62, 70
   Sample session, 24
Save
   STAT101 session, 16, 67
   Worksheet, 24, 62
Scatter plot, see PLOT
SCHOOLS.MTW, 262
Screen colors, changing, 6
Session
   Ending, 187
   Printing, 68, 149
   Sample, 13
   Saving, 67
Session screen
   Accessing from Data Editor, 81
   Clearing, 166
   Definition, 11
   Help, 58
SET, 70, 173
SIGNS, 176
SIN, 176
SINTERVAL median confidence interval, 176
SNOWFALL.MTW, 171, 263
SORT, 177
Spearman's rho, 112
SPEED2.MTW, 263

SQRT square root, 178
SSQ sum of squares, 178
STA261.MTW, 112, 167, 177, 181, 186, 187, 203, 214, 217, 218, 263
STA368.MTW, 264
STACK, 179
Stack worksheets, 66
Standard deviation, 115
Standard error of the mean, 115
Starting STAT101, 10
   From any directory, 6
   Sample session, 15
STAT> prompt, 11, 16, 20, 59
STAT101
   Backing up program disk, 9
   Command file, 223
   Error detection, 221
   Features, vii
   Installing, 4
   Manual, v, 38
   Network, STAT101 not running on one, 8
   Package contents, 1
   Personalizing the program disk, 4
   Running under Windows, 2
   Starting, 10
   Starting from any directory, 6
   Stopping, 11
   Technical Support, 2
STDEV standard deviation, 180
STEALS.MTW, 264
STEM-AND-LEAF, 180
STEPWISE, 107, 142, 182
STEST sign test, 186
STOP, 11, 187
Stopping STAT101, 11
STORE commands in a macro, 70, 187
Stored constants, see Constants
SUBC> prompt, 20, 55, 59
Subcomand, 20, 55
   Sample session, 20
Subset data using logical operators, 93
SUBTRACT, 188, 203
SUM, 188
Symbols, 58
SYSTEM for DOS access, 67, 188

System requirements, 1
TABLE, 189
Tables
  Commands, 79
TALLY, 195
TAN, 197
TBILLS, 264
TECHN.MTW, 265
TEST.MTW, 107, 266
Text file, see ASCII file
Three-dimensional plot, see
  TPLOT
Time series
  Commands, 79
  Plot, see TSPLOT
TINTERVAL mean confidence
  interval, 198
TPLOT three-dimensional plot,
  198
TRANS.MTW, 266
Transformations, Arithmetic, 91
Transforming data
  Commands, 74
TREES.MTW, 215, 267
Trimmed mean using
  DESCRIBE, 114
Troubleshooting, 221
TSPLOT time-series plot, 200
TTEST, 202
  Paired data, 203
TVVIEW.MTW, 267
TWAIN.MTW, 268
TWOSAMPLE, 203
TWOT, 205
  Sample session, 30, 33
TWOWAY, 206
TYPE (DOS command), 147, 208
Typographical conventions, 2, 56

UNSTACK, 209
Uppercase letters

Manual conventions for
  commands, 2
Syntax conventions, 57, 82
USDEMOG.MTW, 268
UTILITY.MTW, 85, 125, 131, 149,
  154, 199, 201, 269

Variables, see Column
Viewing data
  Commands, 72
  Sample session, 19
VOICE.MTW, 269
VPVH.MTW, 270

WALSH, 211
Warranty registration card, 1
WDIFF, 212
WGTHGT.MTW, 147, 271
WIDTH of plots, 213
WINTERVAL, 213
Worksheet, 39
  Adding data, 23
  Clearing, 14, see also RESTART,
    166
  Combining, 65
  Corrections, 23, 46
  Current, 62
  Printing, 69
  Retrieving, 63, 166
  Saving, 62, 172
  Saving to a floppy disk, 63
  Stack, 66
  Storage area, vii
WRITE, 65, 70, 214
  Changing data format, 50
Write-protected disks, 4
WSLOPE, 215
WTEST, 216

ZINTERVAL, 217
ZTEST, 218, 219

STAT101
Software License Agreement

Addison-Wesley License Agreement

READ THIS LICENSE AGREEMENT CAREFULLY *BEFORE* OPENING THIS PACKAGE. BY OPENING THE PACKAGE YOU ACCEPT THE TERMS OF THIS AGREEMENT.

IF YOU DO NOT ACCEPT OR AGREE TO THE TERMS OF THIS AGREEMENT, YOU MAY: (1) RETURN THIS UNOPENED PACKAGE WITHIN 10 DAYS WITH PROOF OF PAYMENT TO THE AUTHORIZED DEALER WHERE YOU TOOK DELIVERY AND (2) GET A FULL REFUND OF THE LICENSE FEE.

Addison-Wesley Publishing Company ("Addison-Wesley") has authorized distribution of this copy of software to you pursuant to a license from Minitab Inc. ("Minitab") and retains the ownership of this copy of software. Minitab retains ownership of the software itself. This copy is *licensed* to you for use under the following conditions.

Permitted Uses/*You MAY*:

*Use the software only for educational purposes.
*Use the software on any compatible computer, provided the software is used on only one computer and by one user at a time.

Prohibited Uses/*You MAY NOT*:

*Use this software for any purposes other than educational purposes.
*Make copies of the documentation, program disk, or backup disk, except as described in the documentation.
*Sell, distribute, rent, sublicense, or lease the software or documentation to any other person, organization, or bookstore.
*Alter, modify, or adapt the software or documentation, including, but not limited limited to, translating, decompiling, disassembling, or creating derivative works.

This License and your right to use the software automatically terminate if you fail to comply with any provision of this License Agreement.

General

Addison-Wesley and Minitab retain all rights not expressly granted. Nothing in this License Agreement constitutes a waiver of Addison-Wesley's or Minitab's rights under the U.S. copyright laws or any other Federal or State law.

Should you have any questions concerning this Agreement, you may contact Addison-Wesley Publishing Company, Inc., by writing to: Addison-Wesley Publishing Company, Inc., STAT101—Marketing Department, Jacob Way, Reading, MA 01867.